Glucose Stability

How Blood Sugar Fluctuations Affect Energy, Insulin, and Metabolic Health

Nathan Reed

PERENNIAL PUBLISHING

Published by Perennial Publishing

Paperback ISBN: 978-1-971996-39-4
Hardcover ISBN: 978-1-971996-40-0
E-book ISBN: 978-1-971996-41-7

Contents

1. Introduction 1
2. Variability Visibility Mechanism 7
3. Spike and Crash Mechanism 15
4. Energy Perception Mechanism 23
5. Input Response Mechanism 29
6. Clearance and Return Mechanism 35
7. Sequencing and Spacing Mechanism 43
8. Rapid Elevation Mechanism 51
9. Reactive Decline Mechanism 59
10. Baseline Drift Mechanism 65
11. Insulin Coupling Mechanism 73
12. Fasting Interaction Mechanism 81
13. Stress Interaction Mechanism 89
14. Pattern Recognition Mechanism 97
15. Threshold Mechanism 105
16. Compensation Mechanism 113
17. Amplitude vs Stability Mechanism 121

18. Integration Mechanism 129

19. Context Dependency Mechanism 137

20. Stability Framing Mechanism 143

Introduction

Glucose is widely recognized because it is visible. It can be measured, tracked, and observed as it changes across time. This visibility creates a sense of familiarity and, with it, a sense of confidence. A number appears, a curve forms, and the signal seems to present its own meaning. Movement in the signal appears to correspond to events, and those events appear to be interpretable in isolation. The result is a form of immediate clarity, where observation is mistaken for understanding.

That clarity does not hold under examination. Visibility does not establish mechanism. A signal can be observed without revealing the processes that produce it. Glucose, despite its accessibility, does not explain itself. It reflects activity within a system that remains largely unobserved, compressing multiple interactions into a single output. What appears to be a direct representation of cause is, in practice, a summary of interaction. The difference between these two is where interpretation begins to shift.

Glucose appears actionable because it changes. Movement suggests that it can be influenced, that each rise or decline corresponds to something that can be adjusted. This creates an illusion of control, where the signal is treated as if it were the mechanism that drives the system rather than the outcome of that system's behavior. Glucose does not initiate or direct the processes that produce its movement. It reflects the combined activity of multiple interacting variables, each contributing to the visible pattern. Treating it as a standalone control point removes the context required to understand it.

The tendency to isolate glucose as a variable arises from a broader pattern of interpretation, where complex systems are reduced to single causes. A change in the signal is linked to a single factor, and that factor is treated as the explanation for the observed behavior. This reduction simplifies interpretation but does so by removing interaction. It assumes that cause can be separated from context, that one mechanism can account for the pattern in its entirety. In a system defined by continuous interaction, this assumption does not hold.

The system in which glucose operates is not a collection of isolated variables but a network of processes that function together across time. Input introduces variability, but its effect depends on timing, composition, and the current state of the system. Hormonal coordination shapes how that variability is distributed and resolved. Clearance determines how movement returns toward baseline. Sequencing organizes how individual cycles interact with one another. Baseline state defines the reference from which all movement begins. Internal signaling introduces variability that is not tied to external input. These processes do not act independently. They operate simultaneously, modifying one another as they unfold.

Glucose is the output of this interaction. It is the observable result of processes that remain interconnected and continuous. The signal does not separate these processes into distinct components. It presents them as a unified pattern, compressing multiple sources of variability into a single curve. This compression is what makes the signal both useful and limited. It provides access to system behavior while concealing the structure that produced it.

Because this compression removes separation between contributing processes, interpretation often fails. Patterns are observed and attributed to singular causes, even when multiple processes contribute to their formation. A rise may be linked entirely to input, ignoring internal signaling or baseline conditions. A decline may be interpreted as resolution, without considering the role of compensation or sequencing.

Stable-looking patterns may be assumed to reflect structural stability, even when underlying processes are actively maintaining their appearance. These interpretations align with what is visible but not with how the system operates.

The same observable pattern can arise from different configurations of the system. A similar rise may reflect different combinations of input, internal signaling, and baseline positioning. A similar decline may result from different clearance dynamics or compensatory responses. The structure of the pattern remains consistent, but the mechanisms that produced it differ. This variability in origin creates ambiguity in interpretation, where the signal alone cannot determine meaning.

Observation, therefore, is insufficient without structure. To interpret the signal, the underlying processes must be organized into a framework that accounts for their interaction. This framework does not simplify the system into a single variable. It defines how multiple mechanisms contribute to the formation of patterns across time. Structure provides the context necessary to move from observation to understanding, linking visible output to the processes that generate it.

This book establishes that structure through a sequence of mechanisms. Each mechanism describes a specific aspect of system behavior, from how variability begins to how it is organized, resolved, and interpreted. These mechanisms do not function in isolation. They interact, overlap, and accumulate across time, forming patterns that reflect the combined behavior of the system. Understanding emerges from how these mechanisms relate to one another, not from any single mechanism alone.

The purpose of this book is to explain how the system behaves. It defines how variability is generated, how patterns are formed, and how signals should be interpreted within the context of interaction. It does not provide instructions, plans, or targets. It does not prescribe actions or define optimal states. It does not simplify the system into a set of steps. Its role is descriptive, not directive. It establishes a framework through which the signal can be understood as part of a larger system.

This distinction defines how the book must be read. Each chapter introduces a mechanism that contributes to the overall structure of the system. These mechanisms build on one another, forming a cumulative understanding that develops across the text. No single chapter provides a complete explanation. Understanding emerges through integration, as each mechanism is placed within the context created by those that precede it. The system operates continuously across time, and the book reflects this continuity by building its explanation in sequence.

Reading the book therefore requires a shift in perspective. It moves away from event-based thinking, where individual changes are interpreted in isolation, toward a pattern-based approach, where behavior is understood across cycles. It replaces immediate reaction with structured interpretation, requiring attention to how patterns form, repeat, and evolve. This shift is not a change in content but a change in how the content is approached. The signal remains the same, but the lens through which it is viewed is altered.

The central shift can be defined in three transitions. Control is replaced by understanding, as the signal is recognized as an output rather than a point of manipulation. Reaction is replaced by interpretation, as patterns are evaluated within the context of system behavior rather than as isolated events. The signal is replaced by the system, as the focus moves from the visible output to the processes that produce it. These transitions establish the framework within which the rest of the book operates.

The signal remains the entry point because it is what can be observed. It provides access to the system, but it does not provide a complete explanation. To understand the signal, the mechanisms that generate it must be defined. These mechanisms describe how variability begins, how it is shaped, and how it is resolved. They provide the structure required to interpret patterns within the system.

The first step in this process is to examine how variability begins. Movement in the signal does not occur without cause. It is initiated by processes that introduce change into the system, establishing the conditions from which all patterns emerge. Before variability can be organized, interpreted, or stabilized, its origin must be understood. The system must be examined at the point where movement begins.

Variability Visibility Mechanism

Glucose becomes observable through variability. This is not a secondary characteristic or a byproduct of system noise, but the primary mechanism through which the signal takes form. At any given moment, glucose exists in motion—rising, falling, stabilizing, or transitioning between states. This movement is continuous, shaped by the interaction of inputs, internal signaling, clearance processes, and the system's prior state. Without this movement, glucose would remain a fixed value without interpretive meaning. Variability is what allows glucose to function as a signal rather than exist as an isolated measurement.

A single glucose value, taken at a single point in time, presents itself as definitive but is structurally incomplete. It captures position without context. It shows magnitude without direction. It provides no indication of whether the system is entering a state, exiting one, or maintaining a stable trajectory. Without movement, the value cannot be interpreted as part of a pattern. It becomes a disconnected observation, detached from the processes that produced it and from the trajectory it is following. The appearance of precision in a single reading masks the absence of temporal information, which is the foundation of signal interpretation.

Meaning emerges through change. The difference between one moment and the next introduces direction. The rate at which that change occurs introduces velocity. The persistence or disruption of that movement over time introduces pattern. These dimensions cannot be derived from a single point. They require continuity. Variability provides that continuity by linking individual moments into a sequence that reveals how the system behaves. Without variability, there is no sequence to observe and no behavior to interpret.

This transformation from value to signal depends on the presence of fluctuation. Glucose does not remain fixed because the system it reflects is not fixed. Input timing introduces periodic increases. Absorption rates determine how quickly those increases occur. Hormonal coordination influences how rapidly glucose is cleared or redistributed. Baseline positioning reflects prior cycles of input and clearance, embedding history into the current state. Endogenous release introduces additional variability independent of external input. Each of these variables contributes to ongoing movement, and their interaction produces the patterns that become visible through variability.

When glucose is viewed through isolated readings, these interactions remain hidden. The system appears simplified, reduced to a sequence of unrelated values. In this framing, variability is often interpreted as inconsistency or instability, not because of its inherent properties, but because the structure needed to interpret it is absent. Without continuity, fluctuation appears random. Without pattern recognition, movement appears chaotic. The absence of temporal framing leads to misinterpretation of the signal, not because the signal lacks structure, but because the structure is not being observed.

When continuity is introduced, variability resolves into pattern. The same fluctuations that appear erratic in isolation begin to show repetition, range, and form when viewed across time. Peaks and declines align with input timing. Rates of increase and decrease reflect absorption and clearance dynamics. Baseline shifts reveal accumulation or depletion across cycles. Variability becomes organized, not because it changes, but because the frame of observation expands to include the relationships between moments.

This shift from isolated readings to continuous patterns alters the interpretation of glucose entirely. Instead of asking what a value is, the focus shifts to what the system is doing. Instead of evaluating position, the emphasis moves to behavior. The signal is no longer defined by where it is at a single point, but by how it moves, how consistently it moves, and how that movement aligns with the broader system context. Variability is the mechanism that enables this transition.

Patterns emerge when variability is structured. Stability, in this context, is not the absence of fluctuation but the presence of consistent pattern formation. A stable system produces variability within predictable ranges, with movements that follow recognizable trajectories. Input leads to increases. Clearance leads to decreases. Baseline levels remain within a defined range across repeated cycles. The pattern persists even as individual values change. Stability is therefore expressed as coherence across time, not as constancy at any single point.

Instability, by contrast, is reflected in the disruption of these patterns. Variability becomes less structured, less predictable, and less aligned with the underlying processes. Peaks may occur without clear relation to input. Declines may be delayed or prolonged. Baseline levels may shift without returning to prior ranges. The signal remains variable, but the pattern loses coherence. This distinction cannot be made through isolated readings. It requires observation of variability across time to identify whether the movement forms a stable pattern or reflects disruption.

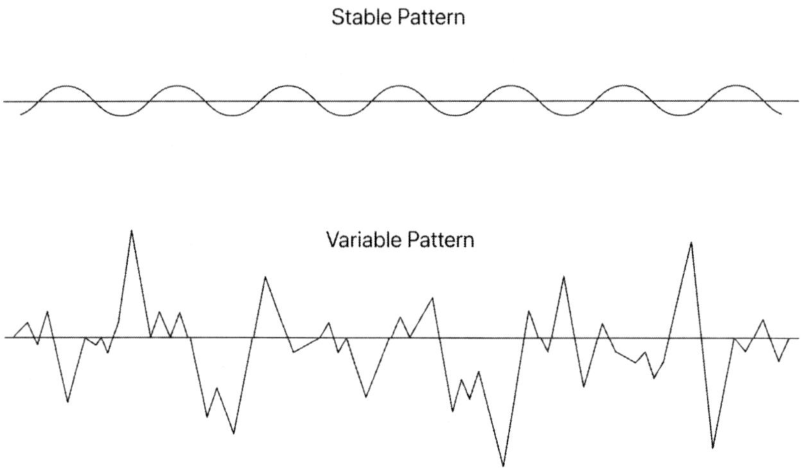

[Figure 1: Stable vs variable glucose line comparison]

The difference between stable and unstable patterns is not defined by the presence or absence of fluctuation, but by the structure of that fluctuation. A stable pattern may include significant movement, but that movement is organized, repeatable, and aligned with system inputs and processes. An unstable pattern may include similar magnitudes of change, but the timing, rate, and recovery differ, producing a pattern that lacks consistency. Variability is present in both cases. What differs is how that variability is structured across time.

The perception of glucose behavior extends beyond measurement. Variability is not only observable through data but is experienced through changes in energy, focus, and physical state. These experiences are not separate from the signal but are expressions of it. As glucose rises, declines, or stabilizes, the system responds in ways that become perceptible. The timing and structure of these responses correspond to the underlying variability. A rapid increase followed by a sharp decline may be experienced differently than a gradual rise and controlled return to baseline. These experiential differences reflect the same patterns observed in measurement, reinforcing that the signal operates across both measurable and perceptual domains.

This alignment between measurable variability and experienced state further emphasizes that glucose cannot be reduced to isolated values. A single reading cannot capture the trajectory that leads into it or the movement that follows. It cannot account for the rate of change that may influence perception or the pattern that may repeat over time. Without variability, the connection between measurement and experience remains fragmented. With variability, the signal integrates across these domains, providing a coherent representation of system behavior.

The limitations of single readings become more apparent when considering what they exclude. Direction is absent, leaving no indication of whether the system is entering or exiting a state. Velocity is absent, removing any understanding of how rapidly the system is changing. Context is absent, disconnecting the value from prior inputs, ongoing processes, and expected outcomes. Continuity is absent, preventing the formation of patterns that would allow interpretation. Each of these dimensions is necessary for understanding the signal, and each is provided by variability across time.

The reliance on isolated values introduces a structural constraint on interpretation. It reduces a dynamic system to a static representation, creating the illusion of clarity while removing the information required for meaningful analysis. Variability restores that information by reintroducing time into the signal. It allows individual values to be connected, compared, and understood as part of a larger pattern. This does not add complexity to the system; it reveals the complexity that is already present but obscured by static observation.

Variability is therefore not a deviation from stability but a prerequisite for its observation. Without fluctuation, there would be no movement to track, no pattern to form, and no behavior to interpret. Stability would exist as an unobservable state, defined only by the absence of change. It is through variability that stability becomes visible, as the system demonstrates its ability to maintain consistent patterns despite ongoing movement. The presence of fluctuation allows the structure of that fluctuation to be evaluated, which is where stability is expressed.

Reframing variability in this way shifts the interpretive approach from reaction to observation. Instead of responding to individual values as isolated events, the focus moves to understanding how those values relate to each other across time. Variability is no longer treated as an error or disturbance, but as the medium through which the signal is conveyed. This reframing does not alter the behavior of glucose; it alters the lens through which that behavior is interpreted.

Within the full system context, glucose variability reflects the coordination of multiple subsystems operating simultaneously. Input timing determines when glucose enters circulation. Absorption rate influences how quickly it rises. Hormonal coordination regulates distribution and clearance. Baseline positioning reflects the cumulative effect of prior cycles. Activity and recovery introduce additional modulation, altering demand and utilization. Each of these variables contributes to the pattern that emerges, and none can be fully understood in isolation. Variability integrates their effects into a single observable signal.

As this integration unfolds across time, patterns become more defined. Repeated cycles of input and clearance establish rhythms. The consistency of these rhythms indicates alignment within the system. Disruptions to these rhythms indicate changes in one or more underlying variables. Variability provides the structure through which these patterns can be identified, compared, and interpreted. Without it, the system would remain opaque, its behavior hidden behind disconnected values.

The transition from viewing glucose as a static value to understanding it as a dynamic signal establishes the foundation for all subsequent analysis.

Variability is the mechanism that enables this transition, revealing how the system behaves rather than simply where it is at a single point. As patterns become more apparent, the signal begins to differentiate into distinct forms, where the structure of variability produces recognizable behaviors that extend beyond general fluctuation into specific patterns of rise and decline.

Spike and Crash Mechanism

The spike and crash pattern emerges as a structured sequence in which rapid elevation is followed by a compensatory decline. This sequence is not incidental. It reflects the coordinated response of a system reacting to the rate and magnitude of change within the glucose signal. Where Chapter 1 established variability as the mechanism that makes glucose observable, this pattern represents one of the first distinct forms that variability takes when examined across time. The signal no longer appears as generalized movement. It begins to organize into recognizable trajectories, where the direction, speed, and resolution of change form a repeatable structure.

A spike begins with an accelerated rise. This rise is defined not only by how high glucose moves, but by how quickly that movement occurs. The rate of increase shapes the system's response as much as the magnitude itself. A gradual elevation distributes the change across time, allowing coordination to unfold with less intensity at any single moment. A rapid elevation concentrates that change, compressing the system's response into a narrower window. This compression alters how signaling unfolds, how quickly clearance mechanisms engage, and how the system transitions from ascent to decline.

As glucose rises, it does so within a system that continuously monitors and responds to change. Hormonal coordination does not operate after the fact; it unfolds in parallel with the rising signal. The elevation itself becomes the input that drives response. The faster and higher the signal moves, the more pronounced the coordination required to manage it. This relationship establishes the second phase of the pattern, where the system transitions from passive observation to active compensation.

The decline that follows is not a separate event. It is directly linked to the preceding rise. The same mechanisms that allow glucose to increase are coupled with mechanisms that regulate its return toward baseline. When elevation is rapid, the compensatory response is often equally pronounced, not as an overcorrection in isolation, but as a continuation of the system's attempt to restore alignment. The decline therefore reflects the system's engagement with the prior elevation. It is not random, and it is not detached from the conditions that produced the spike. It is structurally bound to it.

This coupling between rise and decline forms the core of the spike and crash pattern. The signal does not simply rise and then independently fall. It moves through a coordinated sequence where each phase is dependent on the one before it. The peak represents a transitional point rather than a final state. It is the moment where upward movement gives way to downward correction, where the system shifts from accommodating incoming glucose to redistributing and clearing it. The magnitude and duration of this peak are shaped by both the incoming signal and the responsiveness of the system itself.

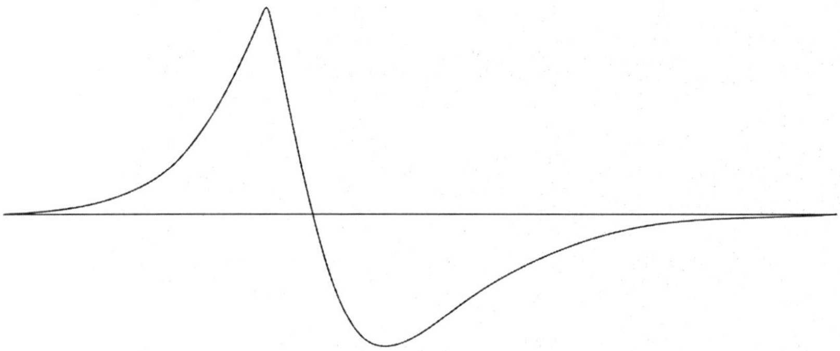

[Figure 2: Classic spike → crash curve]

When viewed across time, this sequence becomes recognizable. The rise accelerates, the peak forms, the decline follows, and the signal returns toward baseline or moves below it before stabilizing again. This full cycle defines the pattern. Observing only the rise isolates the first phase and removes the context required to understand what follows. Observing only the decline disconnects it from the elevation that initiated it. The pattern exists only when the full sequence is considered as a continuous process.

The formation of this pattern depends on the interaction of multiple variables. Input timing determines when glucose enters the system and initiates the rise. Absorption rate influences how quickly that rise occurs, shaping the steepness of the trajectory. Hormonal coordination regulates how the system responds during and after the elevation, influencing both the peak and the subsequent decline. Clearance behavior determines how efficiently glucose is redistributed or removed from circulation. Baseline positioning, shaped by prior cycles, influences where the system begins and how it absorbs new input. None of these variables act independently. Their interaction produces the specific shape of each spike and crash cycle.

Temporal continuity reveals how these variables integrate. The rise is not a singular moment but a progression, where acceleration and deceleration occur within the ascent. The peak is not a fixed plateau but a transition point, where upward momentum slows and reverses. The decline unfolds with its own structure, influenced by the strength and timing of the compensatory response. The return to baseline may occur smoothly or may extend beyond the prior range before stabilizing. Each phase carries information about the system's coordination, and the pattern as a whole reflects how these phases connect.

The compensatory layer is central to understanding why the decline follows the rise. The system does not allow elevation to persist without response. The mechanisms that regulate glucose are designed to maintain alignment across time, and elevation represents a deviation that requires adjustment. When the rise is moderate, the response can be similarly moderate, allowing for a controlled return. When the rise is rapid or pronounced, the response may intensify, producing a steeper decline. This relationship does not imply error or dysfunction. It reflects the system's capacity to respond proportionally to the signal it receives.

The perception of a crash often arises from the experience of this decline phase, particularly when the transition from peak to downward movement is rapid or extends beyond baseline. This experience is not independent of the spike that precedes it. It is part of the same pattern, emerging from the interaction between elevation and response. Without the preceding rise, the decline would not take the same form. The pattern must therefore be interpreted as a unified sequence rather than as isolated events.

Repetition transforms this sequence from a single occurrence into a pattern that shapes system behavior over time. When spike and crash cycles occur repeatedly, they establish a rhythm within the system. The timing, amplitude, and recovery of each cycle influence subsequent cycles, creating continuity across longer timeframes. This repetition can lead to the perception of instability, not because variability is present, but because the pattern becomes more pronounced, more frequent, or less resolved between cycles.

As repetition continues, the system's expectations adjust. Baseline positioning may shift in response to recurring elevations and declines. The timing of responses may adapt to anticipate incoming input. Clearance behavior may change in response to repeated demands. These adjustments do not occur in isolation. They are part of the system's ongoing attempt to maintain coordination across time. The spike and crash pattern therefore extends beyond individual cycles, influencing the broader structure of variability.

Interpretation becomes more complex as patterns repeat. An isolated spike may appear significant when viewed alone, but within a sequence of repeated cycles, its meaning changes. The pattern provides context, showing whether the rise is consistent with prior behavior, whether the decline resolves fully, and how the system transitions between cycles. Without this context, interpretation defaults to reacting to individual peaks or declines, missing the continuity that defines the pattern.

Misinterpretation often arises when only one phase of the pattern is considered. Focusing on the spike without considering the subsequent decline creates an incomplete view of the signal. The elevation appears disconnected from its resolution. Focusing on the crash without recognizing the preceding rise removes the cause from the effect. In both cases, the pattern is fragmented, and the signal loses coherence. Interpretation requires observing the full sequence and understanding how each phase relates to the others.

The spike and crash pattern is not inherently pathological. It is a structural form of variability that emerges under certain conditions defined by rate, magnitude, and system response. Its significance depends on how often it occurs, how pronounced it becomes, and how effectively the system resolves each cycle. A single occurrence does not define instability. It becomes meaningful when viewed within the broader pattern of repetition, where frequency and amplitude begin to shape overall behavior.

Distortion of the pattern occurs when the balance between rise and resolution changes. Repeated rapid elevations may lead to increasingly pronounced compensatory responses, extending the decline phase or delaying return to baseline. Incomplete recovery between cycles may shift the starting point of subsequent rises, altering baseline positioning. Exaggerated declines may produce overshoot below prior ranges, extending the duration of recovery. These distortions do not arise from a single event. They develop through repetition, where the interaction between variables accumulates across cycles.

As distortion increases, the pattern may lose coherence. The timing between phases may become less consistent. The relationship between rise and decline may become less proportional. Recovery may extend across longer periods, reducing the system's ability to reestablish baseline before the next cycle begins. This progression reflects a change in how variability is structured, not the presence of variability itself. The signal remains dynamic, but its pattern becomes less stable.

Reintegrating this pattern into the full system context clarifies its role. Spike and crash represents one expression of variability within a broader range of possible patterns. It is defined by the interaction of input, absorption, signaling, and clearance across time. It does not exist independently of these variables, nor does it define the system in isolation. It is one form that the signal can take when certain conditions align, and its meaning emerges from how it fits within the overall structure of variability.

Understanding this pattern requires maintaining continuity across time, recognizing the link between elevation and decline, and observing how repetition shapes system behavior. The signal does not present itself as isolated events but as connected sequences that evolve with each cycle. As these sequences become more defined, they begin to influence not only measurable behavior but also how the signal is experienced, where the structure of rise and decline translates into shifts in perceived energy and focus that extend the pattern beyond measurement into perception.

Energy Perception Mechanism

Glucose patterns are translated into perceived energy, focus, and fatigue through the structure and rate of change of the signal. This translation is not symbolic or indirect. It is a direct expression of how the system's behavior across time becomes perceptible. Where earlier chapters established that glucose must be interpreted as a continuous signal and that this signal organizes into structured patterns such as spike and crash, the present mechanism extends that framework into the domain of experience. The signal does not remain confined to measurement. It manifests as shifts in perceived state, where movement in glucose becomes movement in energy, clarity, and physical orientation.

This translation begins with the recognition that perception does not arise from static position. A single glucose value, regardless of magnitude, does not inherently produce a defined perceptual state. Instead, perception emerges from how that value is reached and how it changes over time. The system does not respond to position alone; it responds to movement, to transitions, to acceleration and deceleration within the signal. The trajectory leading into a value carries as much relevance as the value itself, and the trajectory that follows continues to shape perception as the signal evolves.

Energy, in this framework, is not a fixed condition but a perceptual output that reflects the structure of glucose behavior across time. Focus follows a similar pattern, aligning with the stability or volatility of the signal as it unfolds. Fatigue emerges not as an isolated state but as an expression of particular patterns of change, especially those involving rapid transitions or extended declines. Each of these perceptual states is therefore linked to the same underlying variable, but differentiated by the structure of its movement rather than by its absolute level.

Rate of change becomes a primary driver of this translation. When glucose rises rapidly, the system experiences not only an increase in availability but also a shift in the pace of change. This acceleration introduces a distinct perceptual shift, one that is different from a gradual elevation reaching the same level. The rapidity of the change alters how the system integrates the signal, producing a perceptual response that corresponds to the speed of transition. Similarly, when glucose declines rapidly, the system does not simply register a lower level. It experiences a transition characterized by downward acceleration, which produces its own distinct perceptual effect.

These transitions define the contour of perception. A gradual rise followed by a controlled decline produces a different experiential profile than a rapid spike followed by a sharp crash, even if both patterns pass through similar ranges of glucose values. The distinction lies in how the signal moves through those values. Perception tracks this movement, not as a conscious interpretation, but as an embedded response to the system's dynamic state. The rate at which glucose changes therefore shapes the intensity, timing, and contrast of perceived energy and fatigue.

This relationship between signal and perception is continuous. As glucose moves through its phases—rising, peaking, declining, and stabilizing—perception follows these transitions in parallel. During the rise, perceptual state may shift in alignment with increasing availability and acceleration. At the peak, perception reflects the transition from upward movement to stabilization or reversal. During the decline, especially when rapid, perception adjusts to the changing availability and the speed of that change. As the signal returns toward baseline or moves below it, perception continues to evolve, reflecting the system's ongoing adjustment.

The spike and crash pattern introduced previously provides a clear example of how this translation occurs. The rapid elevation creates a pronounced upward shift in perception, not because of the value alone, but because of the speed and magnitude of the increase. The subsequent decline, driven by compensatory response, introduces a contrasting perceptual shift, often experienced as a reduction in energy or clarity. These shifts are not independent. They are linked through the structure of the pattern, where the rise sets the conditions for the decline, and the decline reflects the resolution of the rise.

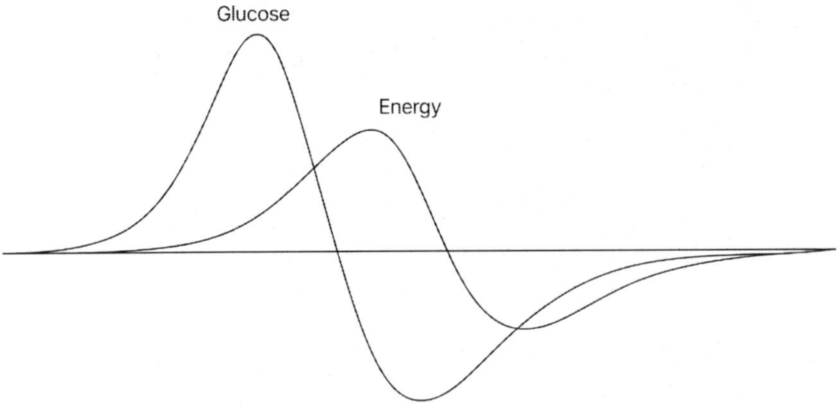

[Figure 3: Overlay of glucose curve and perceived energy curve]

When the glucose curve is overlaid with a corresponding perceptual curve, the alignment becomes visible. The peaks and troughs of perception track the underlying movement of the signal, with variations in timing and amplitude that reflect the system's integration of multiple variables. The perceptual curve does not replicate the glucose curve exactly, but it follows its structure, translating movement into experience. This overlay demonstrates that perception is not arbitrary or detached. It is anchored in the same temporal dynamics that define the signal itself.

Variability amplifies this translation. As the structure of glucose movement becomes more pronounced, the corresponding perceptual shifts become more distinct. Increased variability introduces greater contrast between phases, enhancing the differences between rise and decline, between peak and recovery. This amplification does not alter the fundamental mechanism but increases the visibility of its effects. The more the signal fluctuates, the more perceptible its movement becomes, both in measurement and in experience.

Stability, by contrast, moderates this amplification. When glucose patterns are consistent and movements occur within narrower ranges and at more controlled rates, the corresponding perceptual state becomes more uniform. This does not eliminate variation in perception but reduces its intensity and contrast. The system continues to respond to changes in the signal, but those changes are less abrupt, leading to a more continuous perceptual profile. Stability therefore expresses itself not as a constant state but as reduced volatility in both signal and perception.

This distinction between stability and volatility highlights the role of pattern structure in shaping experience. A system that produces consistent patterns of rise and decline, with controlled rates of change and reliable recovery, generates a perceptual state that aligns with that consistency. A system characterized by rapid fluctuations, repeated spikes, and extended declines produces a perceptual state marked by greater variability and contrast. In both cases, perception reflects the underlying structure of the signal rather than any isolated value within it.

The amplification effect becomes more pronounced when variability increases across repeated cycles. Each cycle contributes to the overall pattern, and as these cycles accumulate, the contrast between phases can intensify. Rapid rises followed by sharp declines create repeated transitions that the system experiences as alternating shifts in state. These shifts may become more noticeable as they recur, not because the system is changing in isolation, but because the pattern of variability becomes more defined across time.

Interpretation of these perceptual shifts requires maintaining the connection between experience and the underlying signal. When perception is viewed independently, without reference to the pattern of glucose behavior, it appears disconnected and potentially inconsistent. A shift in energy or focus may be interpreted as a standalone event, rather than as part of a sequence that includes prior and subsequent changes. This fragmentation leads to misinterpretation, where perception is treated as the primary signal rather than as an output of a deeper system process.

Understanding perception as a translation of glucose patterns restores this connection. It places experiential states within the context of variability, linking them to the structure and rate of change of the signal. This does not reduce perception to measurement, nor does it elevate it as an independent driver. It situates perception within the system, where it reflects ongoing behavior across time. The signal and its perception are therefore not separate domains but different expressions of the same underlying dynamics.

Distortion in perception emerges when the structure of variability becomes less coherent. As patterns become more volatile or less resolved, the corresponding perceptual shifts may become exaggerated or inconsistent. Rapid transitions may produce pronounced peaks and troughs in perceived energy. Extended declines may lead to prolonged periods of reduced clarity or physical engagement. Incomplete recovery between cycles may alter the baseline from which perception begins, creating a sense of instability that reflects the underlying signal.

These distortions do not arise from perception itself. They originate in the structure of the glucose pattern, where repeated variability, amplified transitions, and altered recovery shape the signal over time. Perception follows this structure, translating it into experience. The variability that makes the signal visible also makes its effects perceptible, and when that variability becomes more pronounced or less stable, the perceptual output reflects those changes.

Reintegrating perception into the full system context clarifies its role as an output rather than a driver. The same variables that shape glucose behavior—input timing, absorption rate, hormonal coordination, clearance dynamics, and baseline positioning—also shape the perceptual state by influencing the structure of variability. Changes in any of these variables alter the pattern of the signal, which in turn alters the translation into experience. Perception therefore provides an additional layer of visibility into system behavior, extending the observable signal beyond measurement into lived experience.

This extension does not change the underlying mechanism but expands its domain. Variability remains the foundation of visibility, patterns remain the structure through which variability organizes, and perception becomes the expression of those patterns within the system's experiential framework. The signal is therefore observable in multiple forms, each reflecting the same underlying dynamics across different dimensions.

As the relationship between glucose patterns and perception becomes more defined, attention shifts toward the variables that shape these patterns at their origin. The structure of variability, and therefore the structure of perception, is influenced by how glucose enters and moves through the system, where the timing, composition, and interaction of inputs begin to define the trajectories that ultimately become both measurable patterns and experienced states.

Input Response Mechanism

Glucose movement is initiated and shaped by incoming inputs, with composition, structure, and timing determining the rate and magnitude of the response. This initiation is not a discrete event that begins only when glucose appears in circulation, but a continuous process that unfolds from the moment input is introduced into the system. Where previous chapters established that glucose must be interpreted as a dynamic signal, that this signal organizes into structured patterns, and that those patterns translate into perception, the present mechanism shifts attention upstream to the point where those patterns begin. Variability does not originate at the level of circulating glucose alone. It begins at input, where the characteristics of what enters the system determine how the signal will move across time.

The initiation of glucose movement reflects the transition from absence of input to the introduction of available energy. This transition sets the system into motion. Without input, glucose behavior is shaped primarily by endogenous regulation, maintaining continuity through internal release and baseline positioning. With input, an additional variable enters the system, introducing new material that must be absorbed, distributed, and integrated. This introduction alters the trajectory of the signal, creating the conditions for rise, pattern formation, and subsequent phases of variability.

The rise in glucose does not begin only when absorption is complete. It begins as soon as the process of input is initiated, as early signaling mechanisms prepare the system for incoming change. This preparation reflects anticipatory coordination, where the system does not wait for the full presence of glucose in circulation before responding. Instead, it begins adjusting in advance, aligning hormonal signaling and distribution processes with the expected influx. This anticipatory behavior ensures that the response to input is not delayed but unfolds in parallel with the progression of absorption.

As input is processed, its composition determines how quickly glucose becomes available and how concentrated that availability is across time. Structure, in this context, refers to how the input is organized, how its components interact, and how those interactions influence the rate at which glucose enters circulation. Inputs that are structurally simple or rapidly processed tend to produce a more immediate increase, concentrating the rise into a shorter interval. Inputs that are more complex or integrated with additional components distribute that increase across a longer duration, moderating the rate at which glucose appears in circulation.

This difference in structure shapes both the speed and the magnitude of the response. A rapid input introduces glucose into the system with greater acceleration, producing a steeper rise that may reach a higher peak within a shorter time frame. A more gradual input extends the rise, spreading it across a longer interval and often producing a more moderated peak. These differences do not represent opposing categories but variations along a continuum, where the structure of the input influences how the signal unfolds.

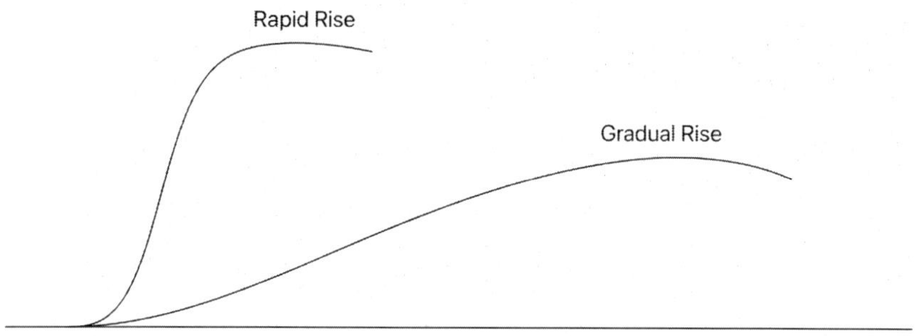

[Figure 4: Fast vs gradual glucose rise from different input types]

The distinction between fast and gradual rise illustrates how input structure translates into observable pattern differences. The glucose curve reflects the timing and concentration of availability, where the steepness of the ascent corresponds to the rate of input processing. This variation in ascent sets the foundation for the subsequent phases of the pattern, influencing how the peak forms and how the system engages in response.

Inputs rarely exist in isolation. In most cases, they are composed of multiple interacting elements that influence each other's processing. Mixed inputs introduce a layer of interaction that alters the trajectory of glucose movement. Components that slow absorption can extend the duration of the rise, reducing the rate of increase while maintaining overall availability. Components that accelerate processing can concentrate the rise, increasing the rate of change. The interaction between these elements produces a composite effect, where the resulting glucose pattern reflects not a single characteristic but the combined influence of multiple variables.

Isolated inputs, by contrast, tend to produce more defined trajectories, where the absence of moderating interactions allows the inherent processing rate to shape the response more directly. This distinction between mixed and isolated inputs is structural rather than evaluative. It describes how the presence or absence of interaction influences the form of the glucose curve, not the desirability of one pattern over another. The system responds to both, integrating their effects into the broader context of variability.

The anticipatory response plays a critical role in shaping how these inputs translate into movement. As input begins, the system adjusts in advance of full absorption, aligning its regulatory processes with the expected trajectory. This alignment influences not only the rise but also the coordination of subsequent phases. The system's response is therefore not purely reactive but partially predictive, incorporating information about the nature of the input into its ongoing regulation.

Temporal continuity reveals how this process unfolds. Input initiates the sequence, introducing material that begins to alter the signal. Early response mechanisms engage, preparing the system for incoming change. As absorption progresses, glucose rises, with the rate and magnitude shaped by the structure of the input. The peak forms as the rate of incoming glucose slows relative to the system's capacity to distribute and clear it. Each phase transitions into the next, creating a continuous trajectory that reflects both the characteristics of the input and the system's response to it.

The patterns that emerge from this process extend beyond the initial rise. The structure of the input influences not only how quickly glucose increases but also how the subsequent pattern develops. A rapid rise may set the conditions for a more pronounced compensatory response, contributing to the spike and crash pattern described previously. A more gradual rise may produce a different trajectory, where the transition into decline occurs with less intensity. These downstream effects illustrate that input structure is not confined to the initiation phase but continues to shape the signal as it moves through its full cycle.

Variability therefore begins at the point of input. The characteristics of what enters the system define the initial conditions of the signal, influencing how it will move, how it will form patterns, and how it will be perceived. This does not isolate input as the sole determinant of glucose behavior. It places input within the network of interacting variables that collectively shape the signal. Input interacts with baseline positioning, which reflects prior cycles and influences how new input is absorbed. It interacts with hormonal coordination, which regulates how the system responds to change. It interacts with clearance behavior, which determines how glucose is redistributed and removed. Each of these variables modifies the effect of input, creating a dynamic interplay that produces the observed patterns.

The distinction between different input structures becomes more apparent when considering how patterns repeat across time. Repeated exposure to similar input structures can produce recurring patterns of variability, where the system's response becomes more consistent with each cycle. This repetition does not imply rigidity but reflects the system's ability to align its behavior with recurring inputs. Variability remains present, but its structure becomes more predictable, allowing patterns to form and stabilize.

Distortion emerges when the characteristics of input interact with other variables in ways that alter the balance of the pattern. Rapid inputs may produce sharper rises that require more pronounced coordination, potentially influencing the subsequent decline. Slower inputs may distribute the rise in a way that alters the timing of peak formation and resolution. These effects are not inherently destabilizing. They become relevant when viewed in the context of repetition, where the accumulation of specific pattern structures influences the overall behavior of the system.

The perception of these patterns, as established previously, reflects the same underlying dynamics. The rate at which glucose rises, determined in part by input structure, translates into perceptual shifts in energy and focus. The timing and resolution of the pattern influence how these perceptual states evolve across time. Input therefore shapes not only the measurable signal but also its experiential expression, linking the upstream variables to downstream perception.

Reintegrating input into the full system context clarifies its role as the initiating layer of variability. It is not an isolated cause but a contributing variable that interacts with the system's existing state. The same input may produce different patterns depending on baseline positioning, prior activity, and ongoing regulation. The system does not respond to input in a uniform way but integrates it within its current configuration, producing variability that reflects both the input and the context in which it occurs.

This integration underscores the importance of viewing input as part of a continuous system rather than as a discrete event. The introduction of glucose into the system sets the process in motion, but the resulting pattern depends on how that input is processed, how the system responds, and how subsequent phases unfold. The signal therefore reflects both the characteristics of the input and the dynamics of the system that receives it.

As attention moves beyond the initiation of glucose movement, the focus shifts toward how the system resolves the elevation introduced by input. The rise sets the trajectory, but the full pattern is defined by how the system distributes, utilizes, and returns glucose toward baseline, where clearance behavior becomes the next defining mechanism in the continuous sequence of metabolic signaling.

Clearance and Return Mechanism

Glucose clearance and return to baseline define how the system resolves elevation and restores alignment across time. This resolution is not a passive decline but an active, coordinated process that unfolds with the same structural complexity as the rise that precedes it. Where input initiates movement and establishes the trajectory of elevation, clearance completes that trajectory by redistributing and utilizing glucose within the system. The decline phase therefore represents not the absence of activity, but the presence of regulation, where multiple subsystems engage to bring the signal back toward its dynamic reference point.

Clearance begins as soon as elevation begins. It does not wait for a peak to form before initiating. As glucose enters circulation and begins to rise, the system simultaneously engages mechanisms that regulate its distribution and use. This overlap between rise and clearance reflects continuous system operation, where opposing processes are not sequential but concurrent. The visible decline that follows the peak is therefore the continuation of an ongoing process rather than the start of a new one. It becomes perceptible when the rate of incoming glucose decreases relative to the system's capacity to distribute and clear it.

This process involves coordination across multiple domains. Glucose is not simply removed from circulation; it is directed into pathways where it is utilized or stored. Cells take up glucose in response to signaling that reflects both current levels and anticipated needs. Tissues vary in their uptake dynamics, contributing to the overall shape of the decline. The liver plays a role in balancing release and storage, adjusting its behavior as the system transitions from elevation to resolution. These coordinated actions collectively shape how the decline unfolds, determining its rate, duration, and completion.

The decline phase therefore reflects an active redistribution of energy within the system. As glucose moves out of circulation and into tissues, the signal changes not because it is disappearing, but because it is being integrated into the system's ongoing processes. This integration requires alignment between signaling, transport, and utilization. When coordination is efficient, the decline proceeds in a manner that returns the signal toward baseline within a defined timeframe. When coordination is less aligned, the decline may extend, altering the shape of the pattern and increasing the duration of variability.

Baseline, in this context, is not a fixed number but a dynamic reference point that reflects the system's current state. It is shaped by prior cycles of input and clearance, by ongoing endogenous activity, and by the broader system context. Return to baseline therefore represents the completion of a pattern, where the system has resolved the elevation introduced by input and reestablished its underlying state. This return does not imply a static endpoint but a transition into the next phase of continuous operation.

The efficiency of clearance plays a central role in determining how stable this pattern appears across time. Efficient clearance is characterized by coordinated movement from peak through decline to return, with a resolution that aligns with the timing and magnitude of the preceding rise. The signal rises, peaks, declines, and returns in a sequence that maintains coherence, allowing each phase to connect smoothly to the next. This coherence expresses itself as stability, not because variability is absent, but because the variability is structured and resolved within consistent bounds.

Efficiency, in this context, reflects the alignment between the processes that govern distribution, utilization, and signaling across time. When these processes operate in coordination, the transition from elevation to return unfolds without disruption, allowing the system to resolve the signal in proportion to the rise that preceded it. This alignment is not defined by speed alone, but by timing and integration, where each component of clearance engages at the appropriate moment relative to the evolving signal. The resulting pattern maintains continuity, with the decline neither lagging behind nor exceeding the requirements set by the prior elevation.

When this alignment shifts, efficiency changes not as a binary condition but as a variation in coordination. Clearance may still occur, but the timing of its components may become less synchronized, altering the shape of the decline and the duration of the return. These shifts do not interrupt the system's operation but modify how resolution unfolds across time. The pattern remains continuous, but its structure becomes extended or uneven, reflecting the degree to which the processes governing clearance remain aligned with the signal they are resolving.

When clearance is less efficient, the pattern extends. The decline may occur more slowly, prolonging the period during which glucose remains elevated. This extension increases the duration of variability, altering how the signal interacts with subsequent inputs and how patterns accumulate across time. The system continues to regulate, but the timing of that regulation shifts, creating a different pattern structure. This difference is not defined by failure but by altered coordination, where the processes involved in clearance operate with different dynamics.

Temporal continuity reveals the full sequence of this process. Following the peak, the rate of incoming glucose decreases, allowing the ongoing processes of distribution and utilization to dominate. The decline unfolds as these processes continue to act, gradually reducing circulating glucose. As the signal approaches baseline, the rate of change slows, transitioning into a phase of stabilization where the system reestablishes its underlying state. This sequence completes the pattern initiated by input, forming a continuous cycle that links rise, peak, decline, and return.

Delayed clearance introduces additional complexity into this sequence. When the decline is prolonged, the system remains in an elevated state for a longer period, extending the pattern beyond its typical duration. This extension increases the interaction between phases, as subsequent inputs may occur before the previous cycle has fully resolved. The overlap between cycles alters baseline positioning, as the system begins the next rise from a different reference point. This shift does not occur abruptly but accumulates across repeated patterns, gradually altering the structure of variability.

The presence of delay also influences how the peak and decline are perceived within the pattern. A prolonged elevation may reduce the contrast between phases, while an extended decline may increase the duration of transition back toward baseline. These changes affect not only the measurable signal but also its integration within the system, influencing how patterns repeat and how stability is expressed over time. The system continues to operate continuously, but the shape of its patterns reflects the altered timing of clearance.

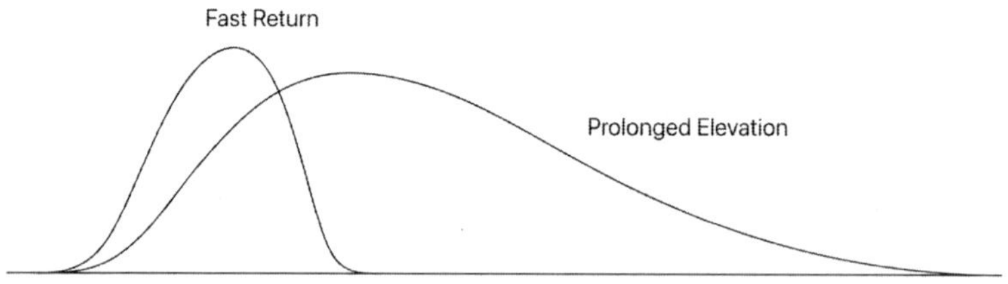

[Figure 5: Normal vs delayed return to baseline]

The comparison between a coordinated return and a delayed return illustrates how efficiency shapes pattern structure. In a coordinated pattern, the decline proceeds at a rate that aligns with the preceding rise, allowing the system to return to baseline within a consistent timeframe. In a delayed pattern, the decline extends, shifting the timing of return and altering the relationship between phases. Both patterns involve the same fundamental processes, but their coordination differs, producing distinct expressions of variability.

Pattern completion depends on this return. The cycle initiated by input is not complete until the system has resolved the elevation and reestablished its baseline. When return is consistent, patterns remain discrete, with each cycle beginning and ending within a defined range. When return is incomplete or delayed, patterns begin to overlap, creating continuity that extends across cycles. This overlap introduces additional variability, as the system integrates multiple phases simultaneously rather than resolving each one fully before the next begins.

The accumulation of delayed clearance across repeated cycles can shift baseline positioning. As the system operates without fully returning to its prior state, the reference point itself changes, influencing how subsequent inputs are processed. This shift is not static but dynamic, reflecting the ongoing interaction between input, clearance, and baseline behavior. As this interaction continues across repeated cycles, the structure of variability begins to reorganize around the new baseline. Each subsequent rise is initiated from a different reference point, altering both the relative magnitude of elevation and the conditions under which clearance operates. The timing between phases may compress or extend depending on how much of the previous cycle remains unresolved, creating patterns that reflect accumulation rather than discrete resolution. This reorganization does not occur abruptly but emerges gradually, as the system integrates the effects of delayed clearance across time.

This gradual shift influences not only the baseline itself but the relationship between rise and decline within each cycle. As patterns overlap, the distinction between phases becomes less defined, and the system begins to operate within a more continuous state of transition rather than clearly separated sequences. The variability remains structured, but its structure reflects a different form of coordination, where resolution is distributed across cycles rather than contained within each one. This redistribution alters how stability is expressed, not by removing variability, but by changing how that variability is organized across time.

The system adapts to this new context, producing patterns that reflect both current inputs and the accumulated effects of prior cycles.

This accumulation also influences the amplitude and timing of variability. As baseline shifts, the magnitude of subsequent rises may change relative to the new reference point. The timing of peaks and declines may adjust as the system integrates overlapping phases. These changes do not occur in isolation but as part of a continuous process where each cycle influences the next. The structure of variability therefore evolves across time, shaped by the interaction of input and clearance within the system.

Reintegrating clearance into the full system context highlights its role as the resolution phase of variability. Input initiates movement, pattern formation organizes that movement, perception reflects it, and clearance resolves it. Each phase is connected, forming a continuous loop where the output of one phase becomes the input for the next. Clearance does not end the signal but transitions it into a state from which new patterns can emerge.

The relationship between rise and clearance is therefore inseparable. The characteristics of the rise influence how clearance unfolds, and the efficiency of clearance influences how future rises are processed. This interdependence reinforces the need to view glucose behavior as a continuous system rather than as discrete events. Each phase carries information about the others, and the pattern as a whole reflects their integration across time.

As attention extends beyond individual cycles, the timing of input and clearance becomes increasingly relevant in shaping overall pattern structure. The spacing between cycles, the overlap between phases, and the alignment of input with resolution all contribute to how variability organizes across longer timeframes, where sequencing and spacing emerge as the next mechanism defining how patterns persist and evolve within the system.

Sequencing and Spacing Mechanism

The sequencing and spacing of inputs determine how multiple glucose cycles interact across time, shaping the structure, continuity, and evolution of variability. This mechanism operates at a level beyond individual rise and decline, organizing how repeated instances of input and clearance combine into a continuous field of activity. Glucose behavior does not reset between events. It carries forward, with each cycle entering a system already influenced by prior movement. Sequencing therefore defines not just when glucose rises, but how each rise is positioned relative to what has not yet fully resolved.

A single rise followed by decline provides a simplified representation of glucose behavior, but this representation remains incomplete when removed from temporal continuity. In practice, the system operates as a sequence of interacting cycles, where each new input enters into a dynamic state shaped by ongoing clearance, prior elevations, and baseline positioning that reflects accumulated history. The concept of a discrete cycle becomes less defined as multiple inputs are placed across time. Instead of independent events, the system exhibits overlapping trajectories, where the rise of one input intersects with the decline of another, forming a continuous structure of variability.

The stacking effect emerges when inputs occur before prior cycles have completed their resolution. In this condition, a new rise begins while the system is still engaged in clearing a previous elevation. The starting point of the new cycle is therefore not baseline, but an already elevated position that carries residual signal from prior input. This structural overlap alters amplitude across cycles, as each subsequent rise adds onto the remaining elevation rather than beginning from a neutral reference. Peaks no longer reflect the magnitude of a single input alone but the cumulative effect of multiple inputs interacting within the same temporal window.

As stacking continues, the duration of elevation extends. Instead of a rise followed by a clear decline and return, the system maintains a sustained level of increased glucose, with fluctuations occurring within an elevated range. The distinction between individual peaks becomes less defined, as successive rises merge into broader structures. The signal appears less segmented, not because variability is reduced, but because variability is distributed across a longer, continuous interval. Transitions between phases lose clarity, as the system moves through overlapping processes rather than discrete sequences.

The decline phase within stacked conditions reflects this integration. It does not represent the resolution of a single rise, but the combined effect of multiple clearance processes operating simultaneously. Each input contributes to the overall elevation, and clearance must address all of them within the same timeframe. The resulting decline is therefore shaped by the interaction of these processes, producing a trajectory that differs from that of an isolated cycle. This difference is structural, emerging from how timing organizes multiple inputs into a composite pattern.

Spacing introduces a different organizational structure. When sufficient time exists between inputs, the system is able to complete the clearance phase and return toward baseline before the next rise begins. This separation preserves the integrity of each cycle, allowing rise, peak, decline, and return to unfold as a coherent sequence. The resulting pattern is composed of discrete units of variability, where each cycle begins from a relatively stable reference point and resolves before the next begins.

This preservation of separation maintains baseline clarity. The system's reference point remains more consistent across cycles, as each return reinforces the underlying state before the introduction of new input. The amplitude of each rise reflects the characteristics of that specific input relative to baseline, rather than being influenced by residual elevation from prior cycles. Pattern repeatability becomes more apparent, as similar inputs produce similar trajectories when starting conditions are consistent. Variability remains present, but its structure becomes more defined, with clear transitions between phases.

The distinction between stacking and spacing therefore lies in how sequencing organizes the interaction between cycles. Stacking compresses separation, combining multiple rises and declines into an extended pattern where phases overlap and integrate. Spacing preserves separation, allowing cycles to remain distinct and fully expressed. Both structures arise from the same underlying mechanisms of input, response, and clearance. The difference is not in what the system does, but in how timing arranges those processes across time.

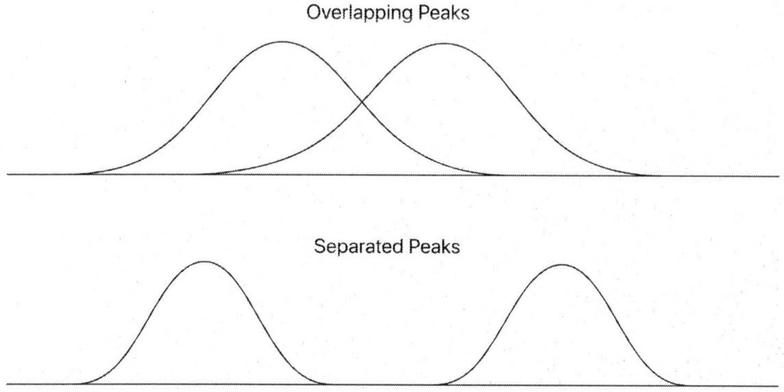

[Figure 6: Stacked vs spaced glucose responses]

The contrast between stacked and spaced responses illustrates how sequencing alters the overall curve structure. In a stacked configuration, the curve extends across a broader time range, with peaks merging and declines integrating into a sustained elevation. In a spaced configuration, the curve shows repeated, separated rises and returns, with each cycle maintaining its own trajectory. These patterns are not static representations but reflections of how multiple cycles interact when placed within different temporal relationships.

Sequencing influences not only the interaction between cycles but also the shape of the combined curve. The ordering of inputs determines how each rise builds upon or separates from prior movement. Inputs that occur in close succession produce cumulative amplitude, where the height of the curve reflects layered contributions. Inputs that occur with greater spacing produce amplitude that resets relative to baseline, maintaining proportional relationships within each cycle. The duration of elevation, the steepness of transitions, and the clarity of phases all depend on how inputs are arranged across time.

Temporal continuity within this framework becomes a continuous sequencing field rather than a series of discrete events. The system operates without interruption, integrating input and clearance across overlapping timeframes. Each moment reflects the combined influence of current input, ongoing clearance, and residual effects from prior cycles. The concept of a beginning and end becomes less distinct, replaced by a continuous flow where phases transition into one another without clear boundaries. Variability is therefore expressed as an evolving structure rather than a repeating loop.

This continuous field integrates multiple variables simultaneously. Input structure influences the rate and magnitude of each rise. Clearance efficiency determines how quickly elevation resolves and how much residual signal remains. Baseline positioning reflects accumulated history, shaping how new inputs are absorbed. Perception translates these combined dynamics into experiential states, reflecting the structure of variability across time. Sequencing acts as the organizing principle that determines how these variables interact, linking them into a coherent pattern.

The perceptual layer mirrors these structural differences. In stacked patterns, perception tends to become more sustained and blended, as the system remains within overlapping phases of elevation and decline. Distinct transitions between states become less pronounced, replaced by a continuous experience that reflects the integrated signal. In spaced patterns, perception aligns more closely with discrete transitions, where changes in energy, focus, and fatigue correspond to the rise and resolution of individual cycles. These differences do not arise from perception itself but from the underlying organization of the signal.

Distortion emerges when stacking persists across repeated cycles without sufficient resolution. As separation between patterns compresses, the system operates within a narrower range of transitions, reducing the opportunity for full return to baseline. This compression leads to reduced resolution opportunity, where each cycle begins before the prior one has completed. The signal becomes continuously elevated, with variability occurring within a constrained range rather than across distinct phases. This does not eliminate variability but alters its structure, making it more complex and less segmented.

The accumulation of stacking also introduces changes in baseline positioning. As the system spends less time at its prior reference point, that reference begins to shift. Subsequent inputs are processed relative to this altered baseline, influencing both the amplitude of rises and the timing of declines. This shift is gradual, emerging from repeated interaction rather than a single event. The system adapts to this new context, producing patterns that reflect both current input and accumulated history.

This accumulation increases the complexity of variability rather than simply its magnitude. The signal becomes composed of multiple overlapping processes, each contributing to the overall pattern. Peaks and declines no longer represent singular phases but integrated outcomes of several interacting cycles. The structure of variability therefore becomes more intricate, with relationships between phases becoming less linear and more interdependent.

Long-horizon structure emerges from this continuous sequencing. Over extended timeframes, the arrangement of inputs and the degree of overlap between cycles shape the overall organization of the signal. Patterns may stabilize into consistent rhythms when sequencing remains relatively uniform, or they may evolve into more complex structures when timing varies. Baseline may remain relatively stable when cycles resolve fully, or it may drift when resolution is consistently incomplete. These outcomes are not fixed but reflect how sequencing organizes variability across time.

Rhythm formation arises when patterns repeat with consistent sequencing, producing predictable structures of rise and decline. This rhythm is not imposed but emerges from the alignment of input timing, clearance dynamics, and system context. When sequencing varies, rhythm becomes less defined, and the signal reflects a broader range of interactions. In both cases, the system continues to operate continuously, with variability structured by how cycles are arranged rather than by any single input.

Viewing glucose behavior as a continuous sequencing field shifts interpretation from isolated cycles to integrated patterns that evolve across time. Each moment reflects the interaction of multiple processes, with sequencing determining how these processes combine. The system does not reset between inputs but carries forward the effects of prior activity, integrating them into ongoing behavior. Stability and instability therefore emerge from how well sequencing allows for coherent interaction between cycles, maintaining structure within the continuous flow of variability.

Reintegrating sequencing into the full system context highlights its role as the structural organizer of variability. Input initiates movement, pattern formation shapes that movement, clearance resolves it, and sequencing determines how these processes connect across time. Each layer builds upon the others, forming a system where variability is continuously generated, structured, and reorganized. Sequencing does not alter the underlying mechanisms but defines how they interact within the temporal dimension.

As sequencing organizes multiple cycles into a continuous pattern, differences in how quickly and how strongly glucose rises within each cycle begin to exert a more pronounced influence on the overall structure. Variations in rate and magnitude introduce new distinctions within the signal, where the speed and scale of elevation become defining features of pattern behavior, leading into the mechanism that governs rapid elevation and its role in shaping glucose dynamics across time.

Rapid Elevation Mechanism

Rapid elevation is defined by variation in the speed and magnitude of upward glucose movement, where the signal does not merely rise, but does so with a particular velocity and scale that shapes the structure of the pattern that follows. The presence of elevation alone does not determine the character of the signal. It is the manner in which that elevation unfolds—how quickly it occurs and how far it extends—that establishes its role within the system. This mechanism introduces a distinction between different forms of rise, where the trajectory itself becomes a defining variable in pattern formation.

The rate of rise, or velocity, determines how quickly glucose transitions from one state to another. A gradual rise distributes this transition across time, allowing the system to engage in coordination while the signal is still unfolding. The change is extended, and the system's response aligns with that extended progression. A rapid rise compresses this transition into a shorter interval, concentrating change within a narrower window. This compression alters how the system experiences and responds to the signal, not because the final level is necessarily different, but because the path to that level is accelerated.

Velocity therefore changes the conditions under which the system operates during elevation. A slower rise provides more continuity between phases, where the distinction between ascent, peak, and early decline may be less pronounced. A faster rise introduces sharper transitions, where the system moves more abruptly from one phase to another. These differences are not secondary characteristics but structural elements that define how the signal behaves across time.

Amplitude extends this distinction by defining the scale of elevation. Where velocity determines how quickly the signal rises, amplitude determines how far it rises. A higher amplitude produces a greater peak, increasing the total exposure of the system to elevated glucose levels. This increased exposure is not limited to the peak itself but extends across the duration of the pattern, influencing how long the system remains within elevated states and how it transitions back toward baseline.

Higher amplitude alters the shape of the entire cycle. As the peak increases, the system engages in a broader range of activity to accommodate and resolve that elevation. The duration of the pattern often extends, as the system requires additional time to redistribute and utilize the available glucose. The decline phase reflects this extension, as the return toward baseline unfolds in response to a larger deviation from the starting point. The relationship between amplitude and duration therefore becomes a defining feature of the pattern, where larger rises are associated with more extended cycles.

The system's response to elevation is coupled to both velocity and amplitude. A rapid, high-amplitude rise introduces a signal that changes quickly and extends significantly, requiring a correspondingly coordinated response. This response is not independent of the rise but is structurally linked to it. The same variables that produce the elevation—input characteristics, absorption dynamics, and system context—also shape how the system engages in its corrective processes. The decline that follows is therefore a continuation of the same pattern, reflecting the system's engagement with the preceding rise.

This coupling establishes proportionality between rise and response. A steeper and larger rise tends to produce a more pronounced corrective phase, not as an overreaction but as an alignment with the magnitude of change that has occurred. The system does not operate in discrete stages where one phase ends and another begins. Instead, the processes of elevation and resolution overlap, with response building as elevation unfolds and becoming more visible as the rate of incoming glucose decreases. The decline therefore mirrors the characteristics of the rise, shaped by the same structural variables.

Temporal continuity reveals how rapid elevation reshapes the full sequence of the pattern. The ascent occurs with increased velocity, reaching a higher peak within a shorter interval. The peak itself may be brief, as the system transitions quickly into the decline phase. The decline unfolds with a trajectory that reflects both the height and speed of the preceding rise, extending or steepening depending on the conditions established during ascent. The return to baseline, or toward it, completes the cycle, with the entire sequence reflecting the influence of the initial elevation.

Rapid elevation also influences the internal variability within a single cycle. As the signal rises quickly and reaches a higher amplitude, the contrast between phases becomes more pronounced. The difference between ascent and decline is amplified, creating a pattern with sharper transitions and greater internal variation. This increased contrast does not change the fundamental structure of the cycle but intensifies its expression, making each phase more distinct relative to the others.

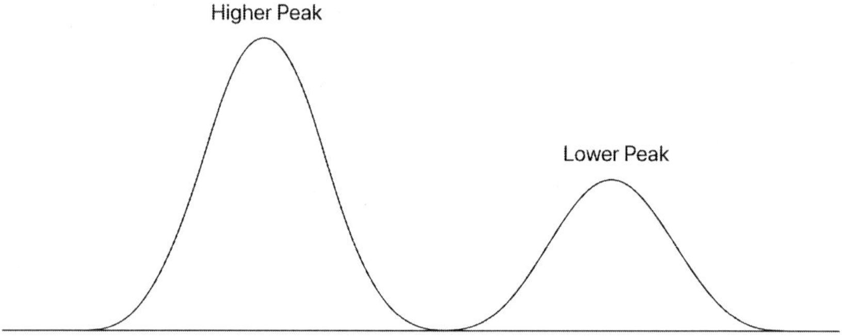

[Figure 7: Low vs high amplitude rise comparison]

The comparison between low and high amplitude rises illustrates how differences in scale and velocity shape the curve. A lower amplitude rise unfolds with a more moderate trajectory, producing a pattern where transitions are less abrupt and the overall range of movement is contained. A higher amplitude rise produces a curve that extends further and changes more rapidly, resulting in a pattern with greater range and sharper transitions. These differences are not categorical but exist along a continuum, where variations in input and system context produce corresponding variations in pattern structure.

The influence of rapid elevation extends into multi-cycle interaction, where amplitude and velocity interact with sequencing. Within stacked sequences, rapid rises contribute to cumulative elevation, increasing the overall height and duration of the combined pattern. The effect of amplitude is compounded, as each new rise builds upon an already elevated state. The resulting curve reflects both the characteristics of individual rises and their interaction across time, producing a structure that differs from isolated cycles.

Within spaced sequences, rapid rises occur from a more consistent baseline, allowing their amplitude to be expressed relative to a stable reference point. The distinction between cycles remains clearer, but the influence of amplitude persists, shaping the height and duration of each individual pattern. The interaction between amplitude and sequencing therefore determines how variability is organized across time, influencing both individual cycles and their integration into longer patterns.

Perception aligns with these structural differences. Rapid elevation produces sharper perceptual shifts, reflecting the accelerated change in the signal. The transition from one state to another occurs more abruptly, creating a more distinct experiential response. Gradual rises produce smoother transitions, where perceptual changes unfold over a longer interval with less contrast between phases. These differences in perception are not independent phenomena but direct translations of the underlying signal structure.

Repeated high-amplitude patterns introduce additional layers of complexity. As larger rises occur across multiple cycles, the system engages in repeated, pronounced responses. The variability within each cycle increases, and the interaction between cycles becomes more complex as amplitude influences both the height and duration of patterns. This repetition amplifies the overall variability of the signal, not simply by increasing its range, but by altering how patterns interact and accumulate across time.

Over-amplification can emerge when high-amplitude rises occur in succession without sufficient resolution. The system operates within a continuously elevated and highly variable state, where each cycle builds upon the previous one. The baseline may shift, reflecting the cumulative effect of repeated elevation and incomplete return. The pattern becomes more complex, with overlapping phases and amplified transitions that reflect the interaction of multiple high-amplitude cycles.

Context determines how rapid elevation functions within the system. The presence of a steep or high-amplitude rise does not inherently define instability or dysfunction. Its significance depends on how it interacts with other variables, including frequency, sequencing, and the system's current state. A rapid rise within a spaced sequence may produce a distinct but contained pattern, while the same rise within a stacked sequence may contribute to cumulative elevation and extended variability. The meaning of amplitude therefore emerges from its placement within the broader system context.

This context dependency prevents reduction of rapid elevation to a singular interpretation. The same structural characteristic can produce different outcomes depending on how it interacts with other elements of the system. Variability, sequencing, and clearance all influence how amplitude expresses itself across time, shaping the resulting pattern in ways that cannot be understood in isolation.

Over longer timeframes, repeated high-amplitude rises contribute to the evolution of variability structure. The system adapts to the patterns it encounters, integrating the effects of repeated elevation into its ongoing behavior. Rhythm may shift as the timing and magnitude of rises influence how cycles align. Baseline positioning may adjust in response to cumulative elevation and resolution patterns. These changes reflect the system's continuous adaptation to the structure of variability it experiences.

Long-horizon effects therefore extend beyond individual cycles, shaping how patterns organize across extended periods. The repetition of high-amplitude rises can lead to patterns that are more pronounced, more complex, or more integrated depending on how they interact with sequencing and clearance. The system does not treat each cycle independently but incorporates the effects of prior patterns into its ongoing operation, producing variability that reflects both current inputs and accumulated history.

Reintegrating rapid elevation into the full system context clarifies its role as a pattern-shaping variable. Input initiates movement, velocity determines the rate of that movement, amplitude defines its scale, and the system's response resolves it through coordinated processes. Sequencing organizes how these cycles interact across time, and perception reflects their structure within the experiential domain. Rapid elevation does not stand apart from these processes but operates within them, influencing how patterns form and evolve.

As attention shifts from the upward phase of the signal to its downward counterpart, the structure of decline begins to reveal its own dynamics, where the speed and depth of reduction introduce additional layers to how patterns are formed and experienced, leading into the mechanism that defines rapid decline and its role within the continuous behavior of the glucose system.

Reactive Decline Mechanism

Decline emerges as a reactive, regulated continuation of prior elevation, shaped by the magnitude and rate of the rise that precedes it. This mechanism does not introduce a new phase that is independent of what came before. It extends the same process that began with input and unfolded through elevation, carrying forward the conditions established during ascent into the resolution phase. The downward movement of glucose is therefore not a separate event but a continuation of the same system behavior expressed in a different direction.

The Reactive Decline Mechanism begins with this continuity. As glucose rises, the system simultaneously engages processes that will ultimately bring it back toward baseline. These processes are not delayed until the peak is reached. They build in parallel with the rise, responding to the rate and magnitude of change as it unfolds. When the rate of incoming glucose begins to slow relative to the system's capacity to distribute and utilize it, the decline becomes more visible. What appears as a transition from rise to fall is in fact a shift in dominance between concurrent processes, where regulatory activity overtakes input-driven elevation.

This coupling between rise and decline establishes that the downward phase cannot be understood in isolation. The conditions created during elevation determine how the decline will proceed. A rapid, high-amplitude rise produces a different decline trajectory than a gradual, lower-amplitude rise, even if both reach similar levels. The system's response reflects the full structure of the preceding movement, integrating both its speed and scale into the resolution process. The decline is therefore shaped by the same variables that shaped the rise, carried forward through continuous system operation.

The rate of decline introduces another layer of structure within this continuity. Just as rise can occur at different velocities, decline can unfold with varying degrees of steepness. A gradual decline distributes the return toward baseline across time, allowing transitions between phases to remain extended and continuous. A steeper decline compresses this transition, producing a more rapid reduction in circulating glucose within a shorter interval. These differences in rate do not alter the fundamental role of decline but change how the system moves through it, affecting duration, transition clarity, and the internal structure of the pattern.

A steep decline shortens the interval between peak and lower levels, creating a pattern where the transition appears more abrupt. The system moves quickly from one state to another, with less time spent within intermediate ranges. A gradual decline extends this interval, allowing the system to move through intermediate states with greater continuity. The distinction between these forms is not one of function but of structure, where the same processes operate under different temporal conditions.

Amplitude remains coupled to this process. Larger prior rises create greater deviation from baseline, requiring a more substantial regulatory response to restore alignment. This does not imply that the system overcorrects but that the scale of response corresponds to the scale of change that has occurred. A higher amplitude rise introduces a broader range over which decline must occur, often resulting in a longer or more pronounced downward phase. The intensity of decline therefore reflects the magnitude of the preceding elevation, maintaining proportionality within the pattern.

Temporal continuity reveals how these relationships unfold across the full cycle. The rise establishes the trajectory, the peak marks the transition, and the decline carries forward the same dynamics into resolution. The system does not reset at the peak but continues to operate within the conditions created during ascent. The return toward baseline completes this sequence, with the decline representing the path through which the system reestablishes its underlying state. Each phase is connected, forming a continuous pattern where direction changes but the underlying processes remain integrated.

The structure of decline contributes directly to pattern formation within each cycle. As the system moves from peak toward baseline, the rate and shape of decline influence how variability is expressed across time. A steeper decline increases the contrast between phases, making the transition from elevation to lower levels more pronounced. A more gradual decline reduces this contrast, creating a smoother progression between states. These differences alter the internal structure of the pattern, influencing how the cycle is perceived and how it interacts with subsequent cycles.

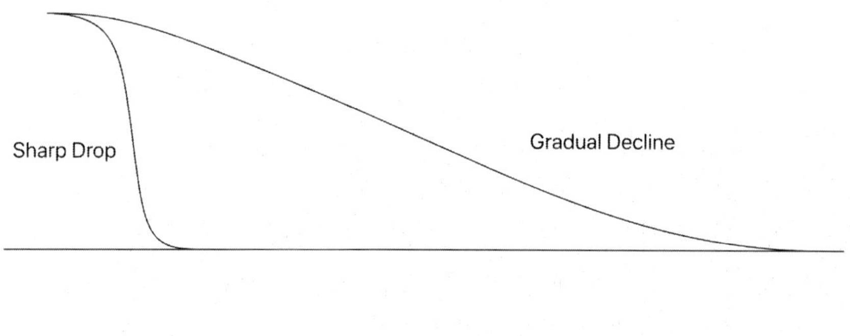

[Figure 8: Sharp vs gradual decline comparison]

The comparison between sharp and gradual decline illustrates how rate shapes the curve. A sharp decline produces a trajectory with a steep downward slope, where the transition from peak to lower levels occurs within a compressed interval. A gradual decline produces a more extended slope, where the transition unfolds over a longer duration. Both represent the same underlying process of regulated return, but their structure differs based on the conditions established during elevation and the system's response to those conditions.

Perception aligns closely with these structural differences. A rapid decline is often experienced as a distinct shift in state, where the speed of change produces a noticeable transition. This perception is sometimes described as a "crash," but within this framework it reflects the rate of decline rather than a separate phenomenon. The system moves quickly from one level to another, and perception tracks that movement. A gradual decline produces a smoother perceptual transition, where changes in state occur more progressively and with less abrupt contrast.

This relationship between rate of decline and perception reinforces that experience follows the structure of the signal rather than its absolute level. The same glucose level may be experienced differently depending on whether it is reached through rapid decline or gradual movement. Perception therefore reflects the dynamics of change, not just the position within the range. The concept of a "crash" becomes a description of how quickly the system transitions rather than an independent event that occurs apart from the pattern.

Within multi-cycle interaction, decline continues to reflect its coupling to prior elevation while interacting with sequencing. In stacked patterns, repeated rapid declines may occur before full resolution is achieved, combining with ongoing elevation from subsequent inputs. The system operates within overlapping phases where decline from one cycle intersects with the rise of another. This interaction alters the shape of both phases, producing composite patterns that reflect multiple simultaneous processes.

In spaced patterns, decline proceeds with greater separation from subsequent rises, allowing the full trajectory from peak to baseline to unfold with less interference. The structure of decline remains linked to the preceding rise, but its expression is more distinct within each cycle. The interaction between decline and sequencing therefore determines how clearly each phase is represented within the overall pattern and how variability organizes across time.

Distortion can emerge when strong declines repeat across multiple cycles, particularly when combined with high-amplitude rises. Each cycle introduces a pronounced contrast between elevation and reduction, increasing the overall variability within the signal. As these cycles accumulate, the system experiences repeated transitions between states with greater intensity, amplifying the differences between phases. This amplification does not arise from decline alone but from the interaction between rise and response across repeated patterns.

Repeated strong declines can also influence baseline positioning over time. As the system moves through pronounced downward phases, the reference point from which subsequent rises begin may shift. This shift reflects the cumulative effect of repeated cycles, where each pattern contributes to the evolving structure of variability. The system adapts to these conditions, integrating the effects of prior patterns into its ongoing behavior.

Context determines how decline functions within this broader structure. A steep decline following a high-amplitude rise within a spaced sequence produces a different pattern than the same decline occurring within a stacked sequence of overlapping cycles. The meaning of the decline cannot be assigned independently of these conditions. It emerges from the interaction between prior elevation, sequencing, and the system's current state. Decline remains a necessary component of the pattern, regardless of its specific form.

Over longer timeframes, repeated decline patterns contribute to the organization of variability across the system. The rhythm of rise and fall, shaped by the rate and magnitude of both phases, influences how patterns align and evolve. Consistent patterns of decline may reinforce certain rhythms, while variation in decline structure may introduce complexity into the overall pattern. These long-horizon effects reflect the accumulation of individual cycles into a continuous field of variability, where each phase contributes to the evolving structure.

Decline therefore participates in shaping long-term pattern evolution, not as an isolated event but as an integrated phase of the cycle. The interaction between rise, peak, and decline defines how variability unfolds across time, with each phase carrying forward the conditions established by the others. The system does not treat decline as a corrective endpoint but as part of an ongoing process that continuously redefines its state.

Reintegrating decline into the full system context clarifies its role as a necessary and structured phase of variability. Input initiates movement, elevation establishes amplitude, response engages in regulation, and decline expresses that regulation as the system moves toward baseline. Clearance completes the process, and sequencing organizes how these cycles interact across time. Decline is therefore not a disruption of the signal but an essential component of its structure, reflecting the system's continuous effort to maintain alignment.

As attention moves beyond the dynamics of decline itself, the focus shifts toward how repeated patterns of rise and fall influence the underlying reference point from which all cycles begin. The interaction between elevation, decline, and incomplete or shifting resolution introduces gradual changes in baseline behavior, where the system's starting position evolves over time, leading into the mechanism that defines baseline drift and its role in shaping long-term glucose patterns.

Baseline Drift Mechanism

Baseline is not a fixed value but a continuously shifting reference point shaped by the accumulation of prior system behavior. The Baseline Drift Mechanism emerges from this continuity, where each cycle of rise, peak, decline, and return contributes incrementally to the position from which future cycles begin. The system does not reset after each sequence. It carries forward the effects of prior activity, embedding them into the conditions that define subsequent responses. Baseline, in this sense, is not an origin but an outcome, formed through repetition and maintained through ongoing interaction.

This framing requires a shift away from viewing baseline as a stable anchor. Instead, it must be understood as a dynamic reference that reflects the integrated history of input, clearance, and sequencing. Each input introduces movement, each clearance phase attempts resolution, and each sequence of cycles organizes how fully that resolution occurs. When these processes repeat across time, they shape the system's underlying position, adjusting the level around which variability unfolds. The baseline therefore becomes a moving coordinate within the system, influenced by the balance between elevation and resolution across cycles.

Drift begins when this balance shifts. Repeated patterns that involve incomplete resolution, overlapping cycles, or sustained elevation gradually alter the baseline. The system spends less time returning to its prior reference point and more time operating within elevated or transitional states. As this pattern repeats, the reference point itself begins to adjust, rising incrementally to reflect the system's new operating conditions. This shift does not occur abruptly or through a single event. It emerges through accumulation, where small deviations from full resolution compound across cycles.

Stacking plays a central role in this process. When inputs occur in close succession, preventing complete clearance between cycles, the system operates within overlapping phases of elevation and decline. Each new rise builds upon residual elevation from prior cycles, reducing the opportunity for return to baseline. Over time, this repeated overlap shifts the system's reference point upward, as the baseline becomes defined not by a fully resolved state but by the average of ongoing elevated activity. The system adapts to this condition, integrating it into its continuous operation.

Amplitude contributes to drift in a complementary way. Repeated high-amplitude rises introduce larger deviations from baseline, requiring more extensive clearance to restore alignment. When these rises occur without sufficient spacing or with incomplete resolution, their effects accumulate. The system remains within elevated ranges for longer durations, and the baseline gradually adjusts to reflect this sustained exposure. The relationship between amplitude and drift is therefore not direct but mediated through repetition and interaction with sequencing and clearance.

Decline, as the reactive phase of the cycle, also influences drift through its completeness. When decline consistently returns the system toward its prior reference point, baseline remains relatively stable. When decline is truncated, delayed, or occurs within overlapping cycles, the return is partial. These partial returns accumulate, contributing to gradual upward movement of the baseline. The system does not interpret this as a discrete change but incorporates it into its ongoing state, adjusting the reference point from which future cycles begin.

The gradual nature of drift makes it difficult to detect within individual cycles. A single rise and decline may appear complete, and the system may appear to return to a stable state. However, when viewed across extended timeframes, small deviations from full resolution become apparent as a pattern. The baseline shifts incrementally, often without noticeable change in any single instance. Only through observation across multiple cycles does the cumulative effect become visible, revealing the slow movement of the system's reference point.

This gradual emergence contributes to the hidden nature of drift. The system may continue to exhibit structured patterns of variability, with recognizable rises and declines, even as the baseline shifts. Locally, within individual cycles, behavior appears consistent and coherent. The structure of variability remains intact, and the system continues to function within its established patterns. However, the position around which these patterns occur has changed, altering the context in which variability is expressed.

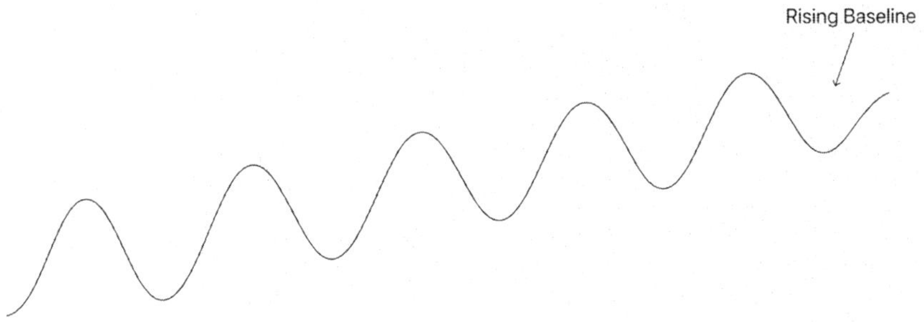

[Figure 9: Baseline shift over time]

The illustration of baseline shift over time reveals this hidden progression. Individual cycles maintain their structure, but the reference line around which they oscillate moves gradually upward. The pattern remains recognizable, yet its position within the system changes. This distinction between local stability and global shift is central to understanding drift. The system can maintain coherence in its immediate behavior while evolving in its underlying state.

Variability continues to operate within this shifted context. Patterns of rise, peak, decline, and return still occur, and the system continues to respond to input and regulate its behavior. However, these patterns now unfold around a different reference level, altering their relative amplitude and duration. A rise that would have occurred from a lower baseline now begins from a higher one, changing its relationship to both the peak and the subsequent decline. The structure of variability remains, but its position within the system has been redefined.

This shift in baseline has implications for system flexibility. As the baseline rises, the range within which the system can operate becomes compressed. The distance between baseline and upper or lower bounds of variability may decrease, reducing the system's capacity to accommodate large fluctuations without altering its structure. Recovery capacity may also be affected, as the system operates within a narrower adaptive space, with less room for full resolution between cycles. This compression does not represent failure but reflects a change in the system's operating conditions.

Flexibility, in this context, refers to the system's ability to move across a range of states while maintaining coherence. A broader range allows for greater variation in input, amplitude, and sequencing without disrupting pattern structure. As baseline rises, this range may narrow, requiring the system to operate within more constrained conditions to maintain stability. The system continues to function, but its capacity to absorb and resolve variation becomes more limited, shaping how future patterns unfold.

As this adaptive range narrows, the system's response to incoming variation begins to change. Inputs that would previously produce a contained and well-resolved pattern may now generate more pronounced or extended responses, not because the input itself has changed, but because the system is operating from a different reference point with reduced capacity for distribution and resolution. The same amplitude and sequencing can therefore produce different pattern structures when baseline has shifted, as the system's ability to absorb and resolve variation becomes more constrained.

This change in responsiveness alters how variability is expressed across time. Transitions between phases may become sharper or more prolonged depending on how the system compensates within its reduced range. Recovery dynamics may extend as clearance operates within a narrower margin, and the interaction between cycles may become more tightly coupled as resolution becomes less complete between inputs. These changes do not introduce new mechanisms but reflect how existing processes behave differently when the system's operating space has been compressed by baseline drift.

The interaction between drift and prior mechanisms reinforces that baseline is an emergent property of system behavior. Sequencing determines how cycles overlap or separate, influencing the degree of resolution between inputs. Amplitude defines the scale of deviation from baseline, shaping the demands placed on clearance. Decline determines how effectively the system returns toward its reference point. Each of these mechanisms contributes to the formation of drift, linking the evolution of baseline to the continuous interaction of input, response, and resolution.

Distortion arises not from any single mechanism but from their accumulation across time. Repeated patterns that favor overlap, sustained elevation, or incomplete return gradually reshape the system's state. This accumulation does not produce an immediate change but builds incrementally, embedding new conditions into the baseline. The system adapts to these conditions, integrating them into its ongoing operation and producing patterns that reflect both current inputs and accumulated history.

Over longer horizons, baseline drift contributes to the evolution of system behavior. The rhythm of variability may change as the timing and amplitude of cycles interact with the new reference point. Patterns that once produced distinct transitions may become more compressed or extended, depending on how the shifted baseline interacts with input and clearance. The system does not revert to a prior state but evolves into a new operating condition, where baseline, variability, and pattern structure are aligned with the accumulated effects of prior cycles.

This evolution highlights that drift exists on a continuum rather than as a binary condition. Baseline may shift incrementally without crossing any defined threshold, and the system may continue to operate with apparent stability within its new context. The meaning of this shift depends on how it interacts with other variables, including the structure of variability, the consistency of patterns, and the system's capacity to maintain alignment across time. Drift therefore represents a change in state rather than a discrete transition from one condition to another.

Reintegrating baseline drift into the full system clarifies its role as an output of repeated behavior. Input initiates movement, elevation establishes amplitude, decline and clearance resolve that movement, and sequencing determines how cycles interact. Drift emerges from the cumulative effect of these processes, reshaping the reference point that defines future behavior. Baseline is therefore not an independent variable but a reflection of how the system has operated over time.

This perspective reframes baseline as a record of system activity, where each cycle contributes to the conditions that define the next. The system carries forward its history, integrating it into its current state and using it as the foundation for future behavior. Drift is the visible expression of this integration, revealing how repeated patterns reshape the system's underlying position without disrupting its immediate coherence.

As the system continues to operate within this evolving state, additional mechanisms begin to emerge that maintain alignment within the new conditions. The system does not passively accept drift but engages in compensatory processes that adjust signaling, distribution, and utilization to maintain functional stability. These processes introduce a new layer of regulation, where the system adapts to its own evolving baseline, leading into the mechanisms that define compensation and its role in sustaining coordinated behavior across time.

Insulin Coupling Mechanism

Glucose and insulin operate as a coupled signaling system, where one provides the observable expression of system behavior and the other coordinates the processes that shape that expression across time. The Insulin Coupling Mechanism establishes that glucose patterns cannot be interpreted independently, because the movement that defines those patterns is continuously regulated by an underlying coordination signal. Glucose appears as the visible trajectory—rising, peaking, declining, and returning—while insulin functions within that trajectory, directing how the system responds to each phase of change.

This coupling begins with the recognition that glucose, as a circulating signal, reflects the outcome of multiple interacting processes, but does not itself direct those processes. Its movement is a consequence of how the system integrates input, distributes available energy, and restores alignment across time. Insulin operates within this integration as a coordinating signal, linking the presence of glucose in circulation to the mechanisms that regulate its uptake, storage, and utilization. The observable signal and the coordinating response are therefore inseparable, each defining the behavior of the other within the same continuous system.

Insulin governs clearance by directing glucose away from circulation and into pathways where it can be used or stored. This role connects directly to the resolution phase described previously, where decline reflects not a passive drop but an active redistribution. As glucose rises, insulin response emerges in parallel, initiating processes that will ultimately determine how the signal returns toward baseline. The rate at which glucose declines, the duration of the pattern, and the completeness of its resolution are therefore shaped by how insulin coordinates this movement across tissues and pathways.

This coordination extends beyond the decline phase into the entire temporal structure of the cycle. Insulin response does not begin after the peak but develops as elevation unfolds, aligning with the rate and magnitude of change. During the rise, insulin activity increases in response to incoming glucose, preparing the system for distribution even as the signal continues to ascend. At the peak, insulin-mediated processes contribute to the transition from ascent to decline, influencing how quickly the system shifts direction. During the decline, insulin continues to guide the return toward baseline, maintaining its role until the pattern resolves. The signal therefore reflects continuous interaction, where insulin is active across all phases rather than confined to a single point.

Pattern structure emerges from this interaction. The rate of decline is not determined by glucose levels alone but by how insulin response aligns with those levels. A coordinated response produces a decline that matches the preceding rise, maintaining proportionality within the pattern. Variations in response timing or magnitude alter this relationship, changing the shape of the curve. The duration of the pattern is similarly influenced, as insulin-mediated processes determine how long glucose remains elevated and how quickly it returns toward baseline. Resolution behavior, including the completeness of return, depends on how effectively insulin coordinates these processes within the system.

As variation in insulin response is introduced, the structure of the glucose pattern begins to diverge in more defined ways. Earlier or more immediate coordination can limit the extent of elevation, compressing the rise and moderating the peak, while delayed engagement allows the signal to extend further before regulation intensifies, producing a higher and more prolonged elevation. Differences in response magnitude further alter this structure, where a stronger coordinating response can accelerate the transition into decline, while a more gradual response extends the duration of elevation before full redistribution occurs. These variations do not create new phases but reshape the trajectory of the existing cycle, altering how the system moves through rise, peak, and resolution.

Across repeated cycles, these differences accumulate into distinct pattern behaviors. Consistently aligned responses tend to reinforce proportional relationships between rise and decline, producing patterns that remain coherent across time. Variability in response timing or magnitude introduces shifts in pattern structure, where elevation, decline, and recovery interact with different degrees of balance. As these interactions repeat, they influence not only individual cycles but the broader organization of variability, contributing to how patterns stabilize, distort, or evolve within the system's continuous operation.

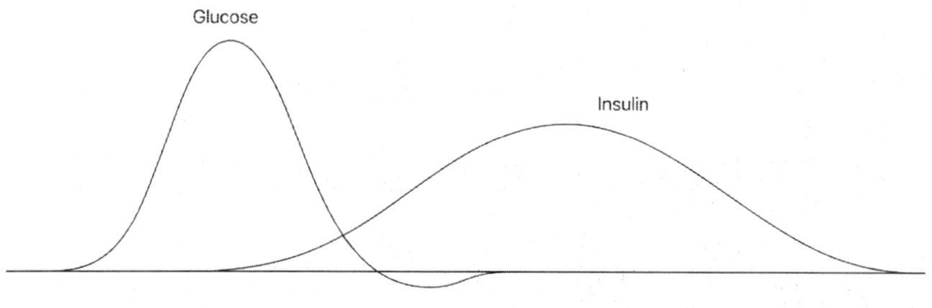

[Figure 10: Glucose curve with insulin response overlay]

Overlaying the glucose curve with the corresponding insulin response reveals this coupling. The rise in glucose is accompanied by an increase in insulin activity, the peak aligns with a transition in response dynamics, and the decline reflects the continued influence of insulin on distribution and utilization. The two curves do not operate independently but move in relation to one another, with timing and magnitude differences that reflect the system's coordination across phases.

Variability in insulin response introduces additional complexity into glucose patterns. The timing of response can shift, with earlier or later engagement influencing how quickly the system begins to regulate incoming glucose. The magnitude of response can vary, altering the strength of the processes that drive clearance. The duration of response can extend or contract, affecting how long regulatory activity persists as glucose returns toward baseline. These variations are not random but reflect differences in system state, input characteristics, and accumulated history. The resulting glucose patterns therefore embody not only the characteristics of input but also the structure of insulin coordination.

When insulin response aligns closely with glucose movement, patterns tend to maintain coherence, with rise and decline occurring within consistent relationships. When this alignment shifts, patterns can become extended, compressed, or otherwise altered in structure. Delayed response may allow glucose to rise further before regulation intensifies, affecting both peak height and subsequent decline. Prolonged response may extend the resolution phase, altering how the system returns toward baseline. These changes reflect differences in coordination rather than the introduction of new mechanisms, reinforcing that glucose behavior remains an expression of coupled system activity.

Dysregulation emerges when the relationship between glucose and insulin response becomes less aligned across time. This does not require extreme deviation but can develop gradually through shifts in timing, magnitude, or duration of response. Delayed clearance may occur when insulin coordination does not match the rate of incoming glucose, extending the elevation phase. Exaggerated rises or declines may appear when response intensity interacts with amplitude in ways that amplify pattern contrast. Pattern distortion, including prolonged cycles or altered baseline positioning, reflects these changes in coordination rather than isolated events.

This layer of dysregulation does not redefine insulin as inherently problematic but highlights its role in shaping how patterns evolve. The system continues to operate through the same coupled mechanism, but the structure of interaction changes, producing different expressions of variability. These expressions remain interpretable within the same framework, where rise, response, and decline are understood as continuous processes influenced by their interaction.

Baseline drift further interacts with insulin behavior by altering the conditions under which response occurs. As the reference point shifts upward through repeated patterns, insulin response operates within a different context, where the system begins each cycle from an elevated state. This change influences how response develops, as the relationship between current glucose levels and prior state affects both timing and magnitude of coordination. Insulin does not act independently of baseline but adapts to it, integrating accumulated system behavior into its ongoing function.

Within multi-cycle interaction, insulin coupling persists across sequences, linking repeated inputs to repeated responses. Each cycle introduces a new instance of coordination, but these instances are not isolated. They overlap and accumulate, just as glucose patterns do, contributing to the overall structure of variability. In stacked sequences, insulin response must address multiple overlapping elevations, integrating its activity across concurrent phases. In spaced sequences, response aligns more distinctly with individual cycles, maintaining clearer separation between phases. The coupling mechanism therefore operates continuously across cycles, shaping how patterns interact and evolve.

Perception reflects this coupling at the level of experience. Changes in energy, focus, and fatigue correspond not only to glucose movement but to the transitions driven by insulin-mediated regulation. Rapid shifts in glucose, influenced by both input and response, produce corresponding perceptual changes. Gradual transitions align with more extended patterns of response. The experiential signal therefore represents the integrated outcome of glucose movement and insulin coordination, translating system behavior into perceptible states.

Over longer horizons, repeated insulin responses contribute to the shaping of pattern stability and system coordination. Consistent alignment between glucose movement and insulin response reinforces coherent patterns, where rise and decline maintain proportional relationships. Variability in response introduces changes that accumulate across cycles, influencing how patterns stabilize or evolve. The system adapts to these conditions, integrating repeated interactions into its ongoing behavior and producing variability that reflects both current inputs and historical patterns.

This chapter introduces insulin as the coordinating layer within the glucose system without extending into detailed analysis of its internal mechanisms. The focus remains on how insulin interacts with glucose to produce observable patterns, establishing the foundation for deeper exploration in a subsequent context. The behavior of insulin itself, including its internal regulation and broader systemic roles, extends beyond the scope of this chapter and will be examined in greater depth within the framework of Insulin Stability.

Reintegrating insulin coupling into the full system clarifies its position within the sequence of mechanisms that define glucose behavior. Input initiates movement, glucose rises with specific velocity and amplitude, insulin response develops in parallel, clearance and decline reflect coordinated regulation, sequencing organizes how cycles interact, and baseline drift reflects accumulated system behavior. Insulin operates within each of these phases, coordinating the processes that produce the observable signal. Glucose remains the visible expression, while insulin provides the structure that shapes its movement across time.

As the system continues to operate through this coupled relationship, the interaction between variables becomes increasingly integrated, where no single mechanism can be isolated from the others. The patterns observed in glucose behavior reflect the combined influence of input, response, clearance, sequencing, and drift, all coordinated through signaling relationships that maintain alignment across time. Understanding this integration requires moving beyond individual mechanisms toward a unified view of system behavior, where each component contributes to the continuous expression of metabolic stability.

Fasting Interaction Mechanism

Fasting interaction begins as a structural change in input conditions, where the absence of incoming energy alters how the system organizes variability across time. The Fasting Interaction Mechanism is defined not by what is added, but by what is removed. Input no longer initiates repeated rises, and the system transitions from externally driven variability to internally sustained movement. This shift does not suspend system activity. It changes how that activity is expressed, redistributing the balance between input-driven cycles and endogenous regulation.

Under conditions of repeated input, glucose behavior is shaped by continuous initiation of new cycles. Each input introduces a rise, engages coordinated response, and moves toward resolution, but subsequent inputs often occur before full clearance is achieved. This produces overlapping sequences, where multiple cycles interact, stack, and reshape the baseline. Variability in this state tends to exhibit higher amplitude, sharper transitions, and more complex multi-cycle integration. The system remains in motion, but that motion is frequently interrupted and redirected by new inputs, creating a pattern structure that reflects both repetition and interference.

In the absence of input, this structure reorganizes. Without new cycles being initiated, existing patterns are allowed to progress toward completion. Clearance processes continue without interruption from additional elevation, enabling the system to move through decline and return phases with greater continuity. The amplitude of variability often decreases, not because the system becomes inactive, but because the absence of new input removes the primary driver of large upward deviations. What remains is a pattern of movement sustained by internal processes, where oscillation persists but with reduced magnitude and fewer abrupt transitions.

This contrast between fed and fasted conditions is not a distinction between activity and inactivity, but between externally driven and internally regulated variability. In the fed state, the system is continually responding to incoming energy, integrating it into existing patterns and reshaping variability with each new cycle. In the fasted state, the system continues to operate, but its behavior reflects endogenous activity rather than repeated external disruption. Variability does not disappear in either state. It reorganizes according to the presence or absence of input.

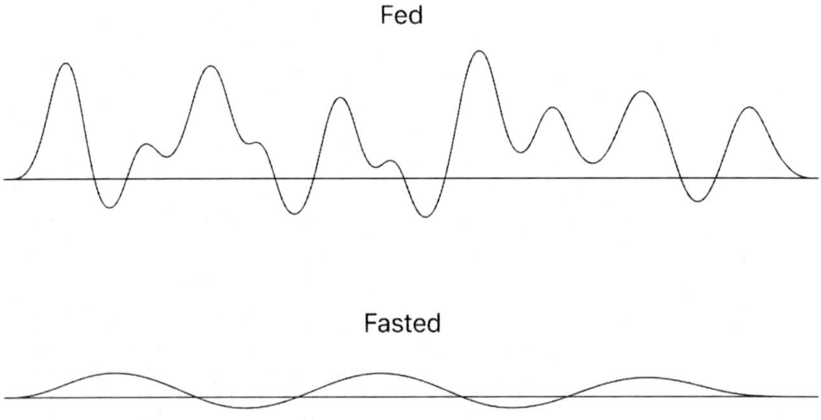

[Figure 11: Fed vs fasted glucose pattern comparison]

In the fed pattern, the curve reflects repeated elevations, overlapping declines, and a baseline that may shift in response to cumulative activity. In the fasted pattern, the curve appears more continuous, with fewer interruptions and more complete transitions between phases. The distinction lies in how variability is structured across time, not in whether it exists.

This reorganization becomes more apparent when considering how variability unfolds across cycles. In the fed state, new inputs often occur before prior cycles have resolved, creating a pattern where phases overlap and transitions compress. Peaks may merge, declines may be truncated, and the system operates within a continuous field of partially resolved activity. In the fasted state, the absence of new input allows each phase to extend toward completion. The decline proceeds without interruption, and the return toward baseline becomes more fully expressed. Patterns become less fragmented, with clearer progression through each phase.

The visibility of baseline increases under these conditions. Without repeated elevation masking the underlying reference point, the system's baseline becomes more directly observable. This does not mean that baseline becomes fixed or simplified, but that its position is less influenced by immediate input. The effects of prior cycles, including any drift that has occurred, are more apparent when not continuously overlaid by new rises. The system's current state is revealed with greater clarity, as the absence of input reduces interference from externally driven variability.

This visibility connects directly to the concept of baseline drift. When drift has occurred, it may remain partially obscured in the fed state by ongoing cycles of elevation and decline. In the fasted state, the system's position becomes more evident, as patterns unfold around the existing baseline without the addition of new peaks. The shift in reference point, accumulated over prior cycles, is not altered by the absence of input but becomes more apparent within the simplified structure of variability.

Clearance completion is central to this transition. In the fed state, clearance often operates within the context of overlapping cycles, where resolution is influenced by new inputs. In the fasted state, clearance can proceed without interruption, allowing the system to complete the process of returning toward baseline. This completion does not represent a final state but a continuation of system operation under different conditions. The separation between phases increases, as each phase is allowed to progress without interference from subsequent cycles.

Despite the absence of input, glucose does not remain static. Endogenous processes maintain movement, ensuring that the signal continues to evolve across time. Internal regulation sustains oscillation, with glucose levels rising and falling within a narrower range defined by the system's current state. This endogenous activity reflects the system's need to maintain continuity, providing a baseline level of variability that persists regardless of external conditions. The pattern therefore shifts from input-driven variability to internally generated oscillation, preserving movement while altering its structure.

Without the interruption of external input, this endogenous oscillation becomes more structurally apparent. The system is no longer required to repeatedly absorb and respond to incoming energy, allowing internally generated movement to define the pattern. These oscillations do not mirror the amplitude or abrupt transitions of input-driven cycles, but they maintain continuity, reflecting the system's underlying regulatory processes. The rise and decline that persist in this state are shaped by internal coordination rather than external introduction, producing a pattern that is more dependent on baseline positioning and system state than on discrete input events. This shift reveals that variability is not dependent on input to exist, but that input changes the scale and structure through which variability is expressed.

As this internally driven pattern continues across time, the system's behavior becomes less segmented and more continuous, with fewer distinct boundaries between phases. Without repeated elevation resetting the trajectory, the system operates within a narrower oscillatory range, allowing longer-term dynamics to emerge with greater clarity. Subtle shifts in baseline, clearance efficiency, and response timing become more visible as they are no longer masked by frequent input-driven disruption. This extended continuity does not simplify the system, but exposes its structure at a different scale, where patterns that develop across multiple cycles can be observed without interference. The system therefore reveals not only how it responds to input, but how it behaves in its absence, providing a more direct view of its underlying organization.

Perception aligns with these changes in variability. In the fed state, higher amplitude and more abrupt transitions produce more distinct perceptual shifts, where changes in energy and focus correspond to the rise and decline of glucose patterns. In the fasted state, reduced amplitude and smoother transitions lead to a more continuous perceptual experience, with fewer abrupt changes between states. This does not imply a uniform or static perception, but a pattern that reflects the underlying structure of variability, where transitions occur with less contrast and greater continuity.

The difference in perception is therefore a direct translation of structural changes in the signal. When variability is organized into overlapping, high-amplitude cycles, perception reflects that complexity and contrast. When variability is reorganized into smoother, lower-amplitude oscillation, perception follows that pattern, producing a different experiential profile. In both cases, perception remains linked to the dynamics of the signal, not to isolated values or static conditions.

This structural shift can be misinterpreted when viewed without context. Reduced variability in the fasted state may be perceived as an improvement or correction, when it is in fact a change in pattern organization resulting from the absence of input. The system appears more stable because the conditions that produce higher amplitude variability are not present. However, this does not necessarily indicate a change in the underlying system state. Baseline, including any drift that has occurred, remains as it was, now more visible but not inherently altered by the absence of input.

Similarly, the exposure of baseline can be misunderstood as a change in the system itself. The absence of repeated elevation reveals the underlying reference point, but does not reset or redefine it. The system continues to operate within its current state, with variability reorganized around that state. This distinction between visibility and change is critical, as it prevents the interpretation of fasting as a mechanism that directly modifies baseline rather than one that reveals it.

The relationship between fed and fasted conditions is therefore not one of opposition but of structural variation within the same continuous system. Input drives variability, introducing repeated cycles that shape pattern formation. Absence of input reorganizes variability, allowing underlying processes to become more visible and cycles to complete without interruption. Both states are part of the system's operation, each revealing different aspects of how variability is structured across time.

Reintegration of fasting interaction into the full system clarifies that it does not exist as a separate mode but as a condition within continuous operation. Input, rise, amplitude, response, decline, clearance, sequencing, and baseline drift all continue to function, but their interaction changes when input is reduced or absent. The system does not switch modes but adjusts its behavior to the conditions present, maintaining continuity while reorganizing variability.

This reorganization expands the interpretive framework, showing that system behavior can be understood not only through the presence of input but through its absence. The patterns observed under fasting conditions provide insight into baseline positioning, clearance efficiency, and endogenous regulation, revealing aspects of the system that may be less visible under continuous input. This insight does not prescribe action but extends understanding, positioning fasting as a structural condition within the broader system rather than a tool or intervention.

As the system continues to operate across varying input conditions, attention shifts toward how control of those conditions influences the structure of variability across longer timeframes. The relationship between input presence, absence, and timing becomes a defining factor in how patterns are organized and maintained, opening the path toward a deeper exploration of how controlled input structures interact with system behavior across extended horizons.

Stress Interaction Mechanism

Glucose elevation can occur in the absence of external input, emerging from internally generated signals that initiate and shape variability across time. The Stress Interaction Mechanism defines this condition, where the system produces upward movement in glucose independent of intake, revealing that variability is not exclusively driven by external energy introduction. The observable signal remains the same—glucose rises, transitions, and resolves—but the origin of that movement shifts from input-driven processes to internally coordinated signaling. This distinction does not create a separate system but introduces an additional layer within the same continuous framework.

The presence of multiple signal sources establishes a dual-driver structure of variability. Input-driven patterns originate from external energy entering the system, following a sequence that begins with intake and progresses through rise, coordinated response, decline, and return. Stress-driven patterns originate internally, without dependence on intake, yet produce comparable observable outcomes. Glucose rises, peaks, and declines under both conditions, creating curves that may appear structurally similar despite differing origins. The system therefore expresses variability through multiple pathways, each capable of generating patterns that converge in their visible form.

This convergence introduces complexity into interpretation. When glucose rises, the visible signal does not immediately distinguish between input-driven and stress-driven origins. The shape of the curve, its amplitude, and its timing may resemble patterns associated with intake, even when no external input has occurred. The system does not label its own output according to source; it integrates multiple drivers into a unified signal. The observer therefore encounters patterns that require contextual understanding to interpret, rather than relying on shape alone.

Internal signaling can operate with sufficient influence to override external conditions. In the absence of input, glucose may still rise, not as an anomaly but as a reflection of system prioritization. The system responds to internal demands by mobilizing available energy, initiating elevation through endogenous processes that sustain function. This does not represent a conflict between internal and external signals but a hierarchy in which internal coordination can supersede the absence of input. The signal therefore reflects the system's immediate requirements rather than the presence or absence of external energy alone.

This hierarchy becomes evident when comparing fed, fasted, and stress-driven conditions. In the fed state, input provides the primary driver of variability, producing repeated cycles of rise and decline that shape pattern structure. In the fasted state, the absence of input allows endogenous processes to maintain continuity, producing lower amplitude oscillations that reflect internal regulation. In the presence of stress-driven signaling, elevation may occur regardless of input conditions, introducing patterns that overlay or interrupt existing cycles. The system does not switch between these states but integrates them, allowing multiple drivers to operate simultaneously.

Patterns formed under stress-driven conditions maintain structural coherence. Glucose may rise rapidly or gradually, reach varying amplitudes, and resolve through decline, following the same fundamental sequence observed in input-driven patterns. The difference lies in the initiating signal rather than the resulting structure. A spike-like pattern may emerge without intake, a sustained elevation may persist independent of recent input, or variability may appear irregular as internal signals fluctuate.

These patterns are real expressions of system behavior, not deviations from it, and must be interpreted within the context of their origin.

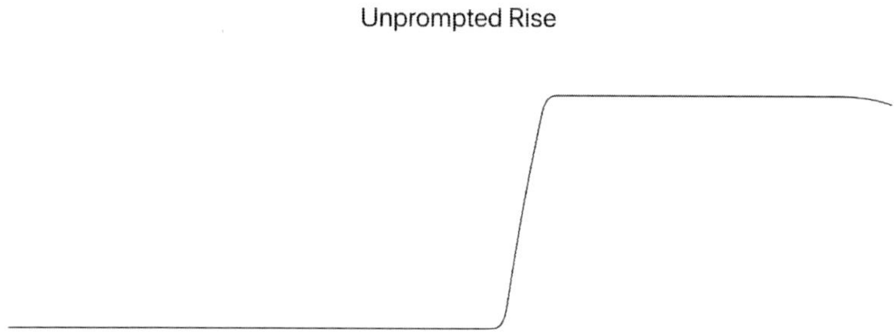

[Figure 12: Stress-induced glucose rise without input]

The illustration of stress-induced elevation demonstrates how glucose can rise in the absence of intake, producing a curve that resembles an input-driven response. The similarity in structure highlights the challenge of distinguishing between sources based solely on the visible signal. The system produces a coherent pattern, but its origin remains embedded within internal signaling rather than external input.

As these internally generated patterns repeat, they participate in the same multi-cycle interactions that define input-driven variability. Stress signals can occur intermittently or persist across extended periods, introducing repeated elevations that interact with existing cycles. When these signals overlap with input-driven patterns, they contribute to stacking, where multiple sources of elevation combine to shape the overall curve. The resulting variability reflects the integration of both drivers, producing patterns that are more complex than those generated by a single source.

This multi-cycle interaction increases the density of variability within the system. Input-driven cycles may be interrupted by stress-driven elevations, altering the timing and shape of the pattern. A decline that would otherwise proceed toward baseline may be interrupted by a new internally generated rise, extending the duration of elevation. Peaks may become less distinct as multiple signals contribute to their formation, and transitions between phases may compress or expand depending on the interaction between drivers. The system continues to operate continuously, but its patterns reflect a higher level of integration.

As stress-driven signaling persists across time, the structure of variability shifts from intermittent disruption to sustained influence. Rather than appearing as isolated elevations, internally generated rises may occur with enough frequency or duration to alter the overall contour of the signal. The system can move into patterns where elevation is not fully resolved before subsequent stress signals emerge, creating extended periods of activity that resemble continuous or semi-continuous elevation. These patterns differ from discrete cycles, as the boundaries between rise, peak, and decline become less defined, and the system operates within an ongoing state shaped by repeated internal signaling rather than isolated events.

This sustained interaction increases the overall density of system activity, where multiple processes operate simultaneously without clear separation. Input-driven cycles, stress-driven elevations, and clearance processes may all be active within the same timeframe, producing a composite signal that reflects layered coordination rather than sequential phases. As this density increases, the system's patterns become more interdependent, with each component influencing the others in real time. The resulting variability is not simply greater in magnitude, but more complex in structure, as the system integrates multiple sources of elevation and resolution into a unified, continuously evolving pattern.

Sequencing becomes less predictable under these conditions. In the absence of stress-driven signals, the timing of input largely determines the structure of cycles, allowing for more consistent sequencing. When internal signals introduce additional variability, they can insert new phases into the sequence, altering the spacing between cycles and disrupting the expected progression from rise to decline to return. This disruption does not eliminate structure but reorganizes it, producing patterns that reflect the combined influence of multiple drivers.

The interaction between stress-driven signals and sequencing connects directly to the principles established in earlier chapters. Stacking occurs not only through closely spaced inputs but also through the addition of internally generated elevations. Spacing may be altered as stress signals compress or extend intervals between phases. Clearance processes continue to operate, but their timing and effectiveness may be influenced by the presence of overlapping signals. The system integrates these factors, producing variability that reflects the complexity of multi-source interaction.

Baseline positioning is also influenced by repeated stress-driven patterns. When internal signals consistently produce elevation, even in the absence of input, the system may spend extended periods above its prior reference point. As with input-driven stacking, incomplete resolution between these internally generated cycles can contribute to gradual upward drift in baseline. This drift occurs without changes in intake, demonstrating that baseline is shaped by overall system behavior rather than by input alone. The system adapts to the combined influence of external and internal drivers, integrating their effects into its ongoing state.

Perception reflects this increased complexity. When glucose variability is driven primarily by input, perceptual changes tend to align with the timing of intake and the structure of associated patterns. Under stress-driven conditions, perceptual shifts may occur independently of input, producing changes in energy, focus, or fatigue that are not directly linked to recent intake. This decoupling increases the unpredictability of experience, as the system's internal signals introduce variability that is not anchored to external events.

The integration of multiple drivers complicates interpretation. Observable patterns may be attributed to input when they originate from internal signaling, leading to misinterpretation of the system's behavior. A rise in glucose may be assumed to follow intake, even when no intake has occurred. Variability may appear inconsistent when viewed solely through the lens of input, as internally generated patterns introduce deviations from expected sequences. The system remains coherent, but the presence of multiple drivers obscures the relationship between cause and effect.

This distortion in interpretation arises from the convergence of pattern shape and variability source. The same structural pattern can emerge from different drivers, making it difficult to assign origin based on appearance alone. The system does not present its signals with explicit markers of source, requiring interpretation to consider context and the possibility of multiple contributing factors. Without this consideration, patterns may be misread, leading to conclusions that do not align with the underlying behavior of the system.

Reintegration of stress interaction into the full system framework clarifies that input-driven, fasting, and stress-driven variability all operate within the same continuous process. Input introduces external energy and initiates cycles, fasting removes that input and reveals endogenous regulation, and stress-driven signaling adds an internal driver that can operate under both conditions. The system does not segregate these influences but integrates them, producing patterns that reflect their combined effect. Variability is therefore a composite expression, shaped by the interaction of multiple sources rather than by any single factor.

This integrated view expands the understanding of glucose behavior beyond a single dimension. Patterns must be interpreted as the outcome of overlapping drivers, each contributing to the structure of variability across time. The system's behavior remains consistent in its mechanisms, but its expression becomes more complex as additional layers of signaling are introduced. This complexity does not indicate disorder but reflects the dynamic nature of a system responding to multiple inputs simultaneously.

As variability becomes increasingly influenced by multiple drivers, the process of interpretation shifts from identifying individual patterns to understanding how those patterns interact within a broader context. The presence of overlapping signals, variable sequencing, and evolving baseline conditions requires a more comprehensive framework for interpreting the signal, where the focus moves from isolated events to integrated pattern recognition across time.

Pattern Recognition Mechanism

Glucose behavior, when observed as isolated events, presents as a sequence of disconnected readings that appear definitive but remain structurally incomplete. A single rise, a single decline, or a single value offers position without context, capturing a moment while excluding the continuity that defines system behavior. The signal exists across time, not within isolated points, and the meaning embedded within that signal emerges only when those points are connected into a sequence. The shift from event-based observation to pattern-based recognition therefore marks a transition from fragmented visibility to structural clarity, where glucose behavior is no longer interpreted through individual moments but through the relationships that persist across repeated cycles.

Patterns form when these relationships repeat with enough consistency to become identifiable. A pattern is not defined by any single phase but by the coherence between phases as they recur across time. The rise, peak, decline, and return are not independent segments but components of a continuous structure, and it is the consistency of their interaction that defines the pattern. When these relationships maintain recognizable proportions—when rises reach similar relative amplitudes, when declines follow comparable trajectories, when sequencing unfolds in a consistent rhythm—the signal begins to exhibit structural repeatability. This repeatability allows the system's behavior to be observed not as a series of unrelated events but as a continuous, organized process.

Recognition operates within this framework as the ability to identify recurring structure without assigning meaning to it. It involves observing how phases connect, how transitions unfold, and how cycles repeat over time. Recognition does not interpret the cause of a pattern, nor does it evaluate its desirability. It identifies the presence of structure, distinguishing between patterns that maintain coherence and those that do not. This distinction is fundamental, as it separates the act of observing from the act of explaining, allowing the signal to be seen clearly before any interpretation is applied.

Within this context, stable patterns are defined by structural consistency across cycles. The amplitude of rises remains within a predictable range relative to the baseline, the timing between phases follows a repeatable sequence, and transitions between rise, peak, decline, and return maintain coherence. Stability does not imply uniformity or absence of variation, but the presence of a consistent relationship between phases that persists across time. Each cycle may differ slightly in magnitude or duration, but the overall structure remains recognizable, allowing the system's behavior to be anticipated within its established pattern.

Unstable patterns, by contrast, are characterized by irregularity in these relationships. Amplitude may vary widely from one cycle to the next, timing may shift unpredictably, and transitions between phases may become distorted or inconsistent. The structure that defines the pattern becomes less coherent, making it more difficult to identify recurring relationships across cycles. This does not imply the absence of variability but a change in how variability is organized. The signal continues to move, but its movement no longer conforms to a repeatable structure, reducing the clarity with which patterns can be recognized.

The distinction between stable and unstable patterns exists along a continuum rather than as a binary classification. Fully stable patterns, where structure is highly consistent across cycles, represent one end of this spectrum. Highly variable patterns, where structure is difficult to identify due to irregularity, represent the other. Between these extremes lie patterns that exhibit partial stability, where some relationships remain consistent while others vary. This continuum reflects the gradual nature of pattern formation and degradation, where changes in structure accumulate over time rather than appearing abruptly.

Patterns rarely transition directly from stable to unstable. The change occurs through progressive degradation of structure, where previously consistent relationships begin to shift in subtle ways across cycles. Amplitude may begin to vary outside its prior range, timing between phases may extend or compress, and transitions that were once smooth may become uneven. These changes do not immediately eliminate the pattern, but they reduce its clarity, moving it along the continuum toward greater irregularity. The structure remains partially recognizable, but its consistency weakens as variation accumulates across repeated cycles.

This progressive degradation alters how patterns can be identified. Early changes may appear as minor deviations within an otherwise stable structure, requiring observation across additional cycles to confirm whether the variation is transient or persistent. As degradation continues, the distinction between phases becomes less defined, and the relationships that previously allowed recognition begin to fragment. The signal does not become random, but its structure becomes more difficult to resolve, requiring broader temporal observation to distinguish underlying patterns from increasing variability.

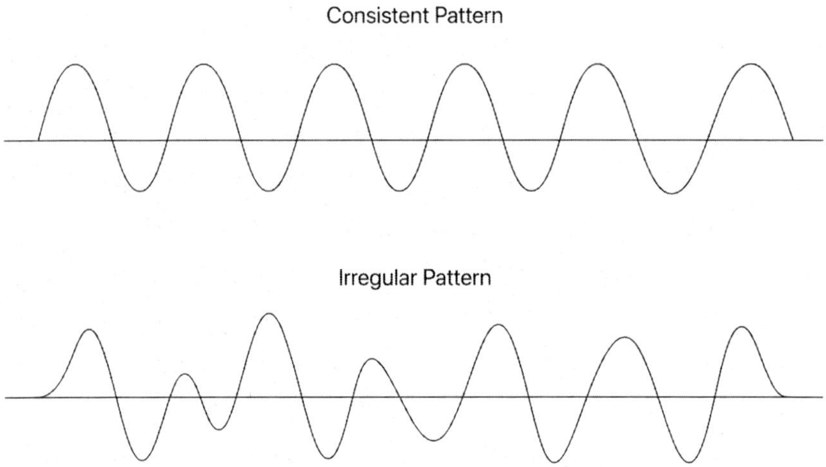

[Figure 13: Stable vs unstable multi-cycle comparison]

The comparison between stable and unstable multi-cycle patterns illustrates how structural coherence manifests across repeated cycles. In the stable pattern, cycles repeat with consistent amplitude, timing, and transitions, forming a recognizable structure that persists over time. In the unstable pattern, cycles vary in amplitude and timing, with transitions that do not align consistently, resulting in a pattern that lacks repeatable structure. The distinction is not based on the presence or absence of variability but on how that variability is organized across cycles.

Patterns become visible only through accumulation. A single cycle may appear similar to another when viewed in isolation, but differences in structure often emerge only when multiple cycles are compared across time. Temporal scaling therefore becomes essential for recognition. Short-term observation may capture individual cycles without revealing their relationships, while longer-term observation allows patterns to emerge through repetition. The ability to recognize structure depends on extending the frame of observation to include sufficient cycles for comparison, allowing consistency or irregularity to become apparent.

Recognition operates independently of interpretation within this expanded frame. It identifies the presence of recurring relationships without assigning cause, value, or implication. A stable pattern is recognized as structurally consistent, but recognition does not determine why that consistency exists or what it implies about the system. Similarly, an unstable pattern is identified through irregularity, but recognition does not assign judgment or prescribe response. This separation ensures that the process of observing remains distinct from the process of explaining, preserving clarity in the initial analysis of the signal.

Distortion in interpretation arises when this separation is not maintained. When isolated events are treated as complete signals, they may be assigned meaning without reference to pattern structure. A single spike may be interpreted as representative of system behavior, despite lacking the context of surrounding cycles. Similarly, a single decline may be viewed as indicative of a trend, without considering whether it aligns with a broader pattern. This event-based interpretation overlooks the continuity of the signal, leading to conclusions that do not reflect the system's ongoing behavior.

The error is not in the observation of the event itself but in the absence of context. Events provide information, but that information remains incomplete until it is integrated into a pattern. Without this integration, interpretation becomes detached from structure, relying on isolated points rather than continuous relationships. The resulting analysis may appear precise but lacks the depth required to capture system behavior across time.

Patterns integrate multiple mechanisms into a unified structure. Sequencing influences how cycles are arranged, affecting the spacing between rises and declines. Amplitude determines the scale of variation within each cycle, shaping the range of movement. Clearance influences how completely each cycle resolves, affecting the transition back toward baseline. Baseline drift shifts the reference point around which cycles occur, altering the relative position of patterns. Each of these mechanisms contributes to the formation of patterns, and recognition requires observing how they interact across cycles to produce coherent or irregular structures.

Perception reflects these structural differences in pattern formation. Stable patterns, with consistent relationships between phases, tend to produce more predictable perceptual experiences, as the system transitions between states in a repeatable manner. Unstable patterns, with irregular transitions and variable amplitude, produce less predictable perceptual changes, as the system's movement does not follow a consistent trajectory. This relationship between pattern structure and perception reinforces that experiential states are linked to the organization of variability rather than to isolated values.

Over longer time horizons, patterns evolve as the system integrates repeated cycles into its ongoing behavior. Consistent patterns reinforce structural coherence, as repetition strengthens the relationships between phases. Irregular patterns introduce variability that may disrupt or reshape existing structures, leading to new patterns emerging over time. This evolution reflects the system's continuous adaptation to its own behavior, where patterns are not fixed but develop through repetition and interaction.

Recognition within this long-horizon context involves identifying how patterns change as well as how they persist. A pattern that appears stable over a short timeframe may exhibit gradual variation when observed across a longer period. Conversely, patterns that appear irregular in the short term may reveal underlying structure when viewed across more cycles. The ability to recognize these shifts depends on maintaining continuity in observation, allowing patterns to be seen as evolving structures rather than static forms.

Recognition operates within limits defined by signal clarity and observation range. When variability remains within structured bounds, patterns can be identified with relative consistency across a moderate number of cycles. As variability increases and structure becomes less defined, recognition requires a longer temporal frame to detect underlying relationships. Short observation windows may capture only fragments of the pattern, making irregular structures appear indistinguishable from noise. Extending observation across additional cycles allows recurring elements to emerge, revealing structure that is not immediately visible within limited segments of the signal.

This distinction between structure and noise becomes central to pattern recognition. Not all variability reflects coherent pattern formation, and not all irregularity indicates the absence of structure. Recognition requires identifying whether variation repeats with underlying consistency or occurs without persistent relationships across time. This determination cannot be made from isolated segments but depends on observing how variability organizes across extended sequences, where structure either stabilizes into identifiable patterns or disperses into non-repeating variation.

Reintegration of pattern recognition into the full system framework clarifies its role as a foundational mechanism for understanding glucose behavior. Input initiates movement, elevation defines the trajectory, insulin coordinates response, clearance resolves the cycle, sequencing organizes interactions, baseline drift shifts the reference point, and internal signals introduce additional variability. Patterns emerge from the interaction of these mechanisms across time, and recognition provides the means to observe that emergence without imposing interpretation.

Pattern recognition therefore operates as a structural lens, allowing the system's behavior to be seen in its full continuity. It identifies the presence or absence of coherence, the degree of variability, and the relationships that define each cycle. It does not assign cause, value, or action, but establishes the foundation upon which interpretation can occur. By separating recognition from interpretation, the system's behavior can be observed clearly, preserving the integrity of the signal before additional layers of analysis are applied.

As patterns become more clearly identified across time, the next stage of understanding involves integrating these observations into a broader framework of system behavior. The distinction between recognizing structure and interpreting its significance introduces a new layer of analysis, where patterns are not only observed but contextualized within the full system, leading into the mechanisms that define how interpretation emerges from recognition within a continuous, multi-variable system.

Threshold Mechanism

Variability is a constant feature of glucose behavior, not an occasional deviation from a stable condition. The signal moves continuously, rising, falling, and reorganizing across time as it reflects the interaction of input, internal signaling, clearance, and sequencing. For this movement to remain coherent, it must occur within bounds that allow the system to maintain structure. These bounds are not externally imposed constraints but emerge from the system's capacity to coordinate its own behavior. The Threshold Mechanism describes how variability operates within these bounds and how crossing them alters the structure, perception, and coordination of the system.

A threshold is not a fixed point or a defined value. It is a boundary condition where the behavior of the system begins to change in a recognizable way. Within this boundary, variability can expand and contract while maintaining coherence, with patterns that remain structurally consistent across cycles. As variability approaches the edge of this boundary, the system begins to transition, not abruptly but progressively, into a different mode of operation. This transition reflects a shift in how the system organizes its movement, not a discrete crossing from one state into another.

Thresholds therefore exist as ranges rather than lines. The system does not encounter a single point where behavior changes instantly, but a zone within which changes begin to emerge. As variability moves through this zone, patterns gradually lose or gain coherence depending on the direction of movement. This gradual transition preserves continuity, ensuring that system behavior evolves rather than resets, even as it moves beyond its typical operating bounds.

Within these bounds, variability maintains a structured relationship between phases. Rises occur with proportional amplitude relative to baseline, declines follow with coordinated timing, and transitions between phases remain coherent. Sequencing allows cycles to unfold with sufficient separation for resolution, and baseline remains a stable reference around which variability occurs. The system is not static within this range, but its movement retains consistency, allowing patterns to repeat in recognizable ways across time.

When variability extends beyond these bounds, the structure of patterns begins to change. This change is not limited to magnitude but affects how phases relate to one another. Transitions may become less defined, with rises extending beyond their typical proportion or declines becoming disproportionate in their response. The sequence of cycles may compress, overlap, or extend, altering how patterns are organized across time. The system continues to operate, but its coordination reflects the altered conditions under which variability is occurring.

These changes can be observed across different structural dimensions. In one case, variability may exceed the system's typical amplitude range, producing elevations that extend beyond the scale at which patterns remain coherent. This amplitude threshold does not represent a limit that should not be crossed, but a point at which the system begins to exhibit different structural relationships between phases. The rise may extend further, the decline may become more pronounced, and the overall pattern may take on a different shape as a result.

In another case, variability may change too rapidly for the system to maintain its usual coordination. When the rate of change accelerates beyond the system's capacity to align response with movement, transitions between phases may become abrupt or less proportional. This rate threshold reflects a change in how quickly the signal moves rather than how far it moves, altering the temporal structure of patterns even when amplitude remains within a familiar range.

Duration introduces a different dimension of threshold behavior. When elevation persists beyond the time frame in which the system typically resolves it, the pattern extends, and the return toward baseline becomes delayed or incomplete. This duration threshold reflects the system's capacity to sustain coordinated behavior over time. As elevation persists, the relationship between phases changes, and the system begins to operate under conditions where resolution is no longer aligned with its typical timing.

Sequencing adds another layer to threshold behavior, where the interaction between cycles determines whether patterns remain distinct or become compressed. When cycles overlap excessively, with new rises occurring before prior declines have resolved, the system moves beyond its typical sequencing structure. This sequencing threshold alters how variability is organized across cycles, creating patterns that reflect cumulative interaction rather than discrete repetition.

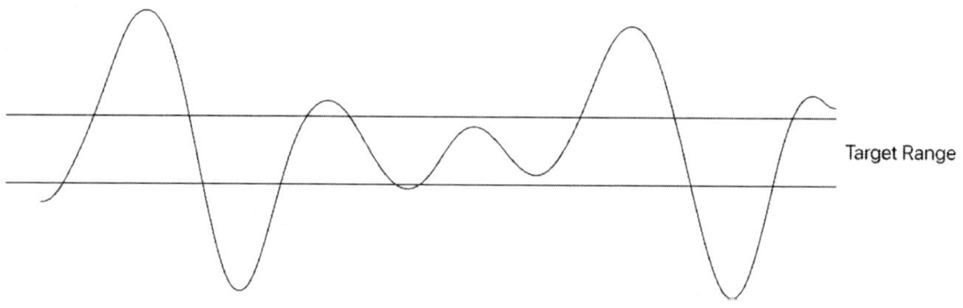

[Figure 14: Threshold band with excursions beyond range]

The threshold band illustrated here represents the range within which variability maintains structural coherence. Movement within this band reflects the system's typical operation, where patterns remain recognizable and transitions remain aligned. Excursions beyond this band do not represent a failure of the system but a shift in how variability is expressed. As the signal moves beyond the central range, patterns become progressively altered, reflecting the system's response to extended amplitude, accelerated rate, prolonged duration, or compressed sequencing.

The position of these thresholds is not fixed but depends on the context within which the system operates. Baseline position influences how much variation can occur before structural relationships begin to change. A system operating around a higher baseline may encounter threshold transitions at different points than one operating around a lower reference, as the relative scale of variability is altered. Sequencing also shapes threshold behavior, as closely spaced cycles reduce the available space for variability to unfold before patterns begin to overlap.

Amplitude interacts with these contextual factors, influencing how quickly the system approaches or moves through threshold zones. Larger deviations from baseline bring the system closer to structural transition, while smaller deviations allow it to remain within its typical range. However, amplitude alone does not define threshold behavior. The system's state, including its prior cycles and accumulated patterns, determines how variability is integrated and where transitions begin to emerge.

As thresholds are approached and crossed repeatedly, the system's behavior begins to adapt. The boundaries within which variability is tolerated may shift, reflecting changes in how the system organizes its movement. In some cases, the range may expand, allowing for greater variation without altering pattern structure. In others, the range may compress, reducing the space within which patterns remain coherent. These changes do not occur instantly but develop gradually as repeated exposure to certain conditions reshapes the system's capacity for coordination.

This progressive adjustment illustrates that thresholds are not static constraints but dynamic features of system behavior. They emerge from the interaction of multiple variables and evolve as those variables interact across time. The system does not operate against fixed limits but within a set of boundaries that reflect its current state and its accumulated history. As that state changes, the boundaries shift, altering how variability is expressed and how patterns are organized.

Perception reflects these threshold dynamics in a way that parallels structural change. Within the threshold range, perception tends to follow the signal with proportional changes, where shifts in energy or focus align with the gradual movement of glucose patterns. As variability moves beyond this range, perception may become more pronounced or abrupt, reflecting the altered structure of the signal. The intensity of experience may increase, transitions may feel sharper, and the relationship between phases may appear less continuous.

These perceptual changes do not define the threshold but illustrate its effect. The system's internal coordination shapes both the signal and its translation into experience, and when coordination changes, perception follows. This relationship reinforces that thresholds operate across multiple layers, influencing not only the structure of patterns but also how those patterns are experienced.

Pattern recognition provides a framework for observing these changes across time. Stable patterns, defined by consistent relationships between phases, tend to remain within threshold bounds, maintaining coherence across cycles. As variability approaches or crosses thresholds, patterns may become less stable, with irregular amplitude, timing, or transitions. This does not assign value to stability or instability but identifies how structural relationships change as variability moves through different ranges.

Threshold crossing can also occur without immediate visibility when viewed within a limited timeframe. A single cycle may not reveal whether variability has extended beyond the system's typical range, particularly if patterns appear coherent in isolation. Only through multi-cycle observation does the accumulation of changes become apparent, revealing shifts in structure that indicate movement beyond threshold bounds. This reinforces the need for continuity in observation, where patterns are evaluated across extended sequences rather than isolated events.

As threshold crossing persists across repeated cycles, the structure of variability begins to reorganize rather than simply deviate. Patterns no longer return to their prior form after each excursion but instead evolve around the conditions created by repeated boundary interaction. The relationships between rise, decline, and return begin to shift, producing cycles that reflect the influence of prior threshold crossings rather than isolated deviations. This reorganization does not eliminate pattern structure but alters it, creating a new form of variability that differs from the system's earlier operating state.

This evolving structure introduces a feedback effect, where each threshold crossing influences the conditions of subsequent cycles. Amplitude may compound across cycles, sequencing may compress further, and clearance may operate within increasingly constrained timing. The system continues to maintain continuity, but its behavior reflects an accumulated response to repeated boundary interaction. Thresholds therefore do not function only as points of transition within individual cycles but as forces that shape how patterns develop across time, linking single excursions into longer-term structural change.

Reintegration of the threshold mechanism into the full system highlights its role as an emergent property of continuous interaction. Input introduces variability, elevation defines its trajectory, insulin coordinates response, clearance resolves movement, sequencing organizes cycles, baseline drift shifts the reference point, and internal signals add additional layers of variability. Thresholds arise from the combined effect of these mechanisms, defining the range within which their interaction maintains coherence and the conditions under which that coherence changes.

Understanding thresholds in this way removes the concept of fixed limits and replaces it with a dynamic model of system boundaries. The system operates within ranges that are shaped by its own behavior, and these ranges evolve as that behavior changes. Variability is not constrained by external definitions but by the system's capacity to coordinate its own movement across time. Thresholds mark the points at which this coordination shifts, revealing how the system adapts to the conditions it encounters.

As variability continues to interact with these dynamic boundaries, the system develops mechanisms to maintain function under changing conditions. These mechanisms do not prevent threshold crossing but respond to it, adjusting coordination to sustain continuity. This transition from boundary behavior to adaptive response introduces the next layer of system function, where compensation emerges as the means by which the system maintains coherence as it operates within and beyond its evolving thresholds.

Compensation Mechanism

Compensation operates as a continuous adjustment process embedded within the system's ongoing behavior, emerging wherever variability begins to challenge coordinated structure. It does not begin at a discrete moment, nor does it activate only under extreme conditions. It is present at all times, modulating how the system responds to change as that change unfolds. When variability shifts the relationships between phases—when amplitude expands, timing compresses, sequencing overlaps, or baseline alignment drifts—the system does not suspend operation. It adjusts in place, preserving continuity through modification rather than interruption.

This adjustment begins with the detection of deviation, which is not an event but a property of continuous monitoring. The system tracks the alignment between phases as part of its normal function, registering how closely rise, peak, decline, and return maintain their established relationships. When these relationships begin to shift, even slightly, the system incorporates that shift into its ongoing feedback. Detection operates across gradients rather than thresholds, identifying emerging divergence before it becomes externally visible. Variations in amplitude, timing, sequencing, and baseline alignment are all integrated into this detection process, forming the basis for subsequent adjustment.

Response adjustment follows as an immediate extension of this detection. The system modifies how it coordinates movement across phases, altering timing to align more closely with changing conditions, adjusting magnitude to balance the scale of variation, and extending or compressing duration to maintain continuity between phases. These modifications reshape how decline unfolds, how clearance resolves elevation, and how sequencing organizes cycles. The goal is not restoration to a prior state but preservation of coordinated behavior within the current context. Adjustment therefore reflects adaptation rather than correction, allowing the system to sustain function as variability evolves.

Through these adjustments, functional continuity is maintained. Patterns may shift in appearance, reflecting altered relationships between phases, but they remain operational. Movement continues, transitions occur, and the signal retains enough coherence to sustain system behavior. The system does not require stable conditions to function. It requires coordinated relationships, and compensation ensures that those relationships persist even as variability extends beyond its typical structure. The pattern that results may differ from prior cycles, but it remains recognizable as a coherent sequence.

This capacity introduces a distinction between visible stability and structural stability. Compensation can maintain the outward appearance of coherent patterns even as underlying conditions change. A sequence of cycles may appear consistent, with recognizable rises and declines, while the processes required to maintain that consistency become increasingly complex. The system may be operating under conditions that require continuous adjustment, yet the observable signal does not immediately reflect this effort. Visible stability, in this sense, can coexist with underlying strain, as compensation masks the internal adjustments required to sustain pattern coherence.

Across cycles, compensation organizes into a feedback loop that integrates detection, adjustment, and temporary stabilization. Within a single cycle, deviation leads to modification, and that modification allows the pattern to proceed with coordinated structure. In repeated cycles, these adjustments do not reset. Each cycle builds upon the modifications introduced by prior adjustments, creating continuity between cycles rather than discrete repetition. The system operates within the context created by its own prior responses, carrying forward the effects of adjustment into future behavior.

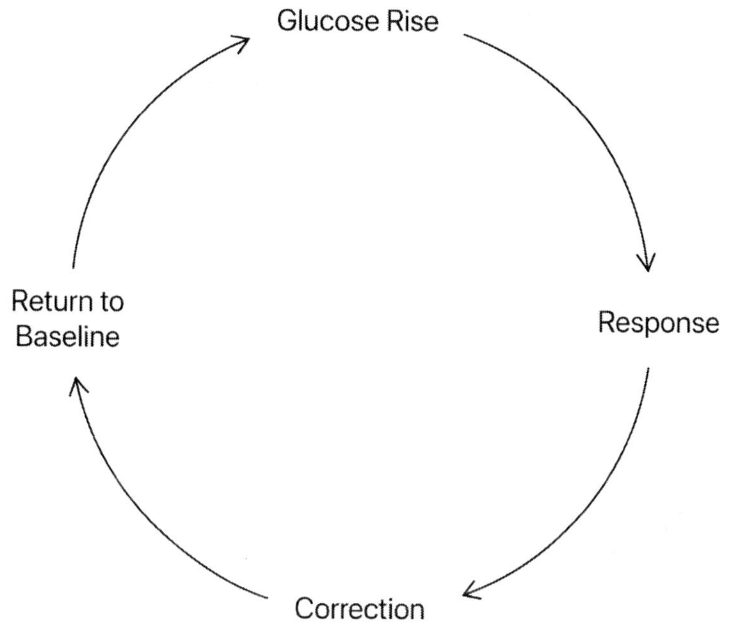

[Figure 15: Compensation loop diagram]

This loop illustrates the progression from deviation to adjustment to stabilization, and then into a new context in which further deviation occurs. The process is continuous, with each phase feeding into the next, forming an ongoing cycle of adaptation that sustains system function across time.

As this loop repeats, adjustment begins to compound. Each modification alters the starting conditions for subsequent cycles, shifting how the system responds to new variability. The system is no longer reacting to isolated deviations but to deviations that occur within an already adjusted state. Responses become layered, reflecting both current conditions and the accumulated effects of prior adjustments. Over time, these layers integrate into the system's baseline behavior, where compensation is no longer a distinct response but part of the ongoing structure of variability.

As these compounded adjustments continue across cycles, the structure of the pattern itself begins to reorganize. The system is no longer maintaining the original relationships between phases through adjustment alone, but gradually establishing new relationships that reflect the accumulated effects of prior compensation. Rises may begin from altered starting positions, declines may follow modified trajectories, and the spacing between cycles may shift as sequencing adapts to the adjusted state. What initially appears as temporary stabilization becomes integrated into the pattern's ongoing form, where compensation no longer preserves the original structure but reshapes it. The system continues to operate coherently, but the coherence now reflects a reorganized pattern that incorporates the history of adjustment rather than returning to its earlier configuration.

This compounding introduces additional demands on coordination. The system must align multiple variables simultaneously, integrating adjustments across amplitude, timing, duration, and sequencing. The coordination load increases, requiring more precise alignment between phases to maintain coherence. As this load grows, the margin within which variability can be managed narrows. The system operates with reduced tolerance for deviation, where small misalignments require immediate adjustment to prevent disruption. At the same time, the system becomes increasingly dependent on the accuracy of its responses. Precision in timing, magnitude, and duration becomes essential for maintaining pattern structure under these conditions.

As compensation continues across cycles, precision begins to shift. Adjustments that initially align closely with deviations may become slightly misaligned. Timing may begin to lag or lead relative to the signal, magnitude may overshoot or undershoot the required response, and duration may extend beyond or fall short of optimal alignment. These changes are subtle and develop gradually, reflecting the cumulative effect of sustained adjustment. The system continues to function, but the precision of its coordination begins to decline.

This stage, where precision drifts without immediate breakdown, represents a transition between stable compensation and progressive weakening. Patterns remain coherent, but their structure begins to show signs of variability that were not previously present. Transitions may become less smooth, amplitude relationships less consistent, and sequencing less predictable. These changes do not disrupt function immediately but indicate that the system is operating under conditions where precise alignment is more difficult to maintain.

As misalignments accumulate, compensation becomes less effective at preserving pattern coherence. Adjustments continue, but their ability to align with deviation diminishes. The system may respond too strongly or too weakly, too early or too late, producing patterns that are increasingly irregular. The decline in coordination does not occur suddenly but develops through the accumulation of small deviations in response. Over time, these deviations alter the structure of patterns, reducing their repeatability and coherence.

At a further stage, compensation approaches saturation. Under these conditions, the system is actively adjusting across all dimensions simultaneously. Every phase requires modification, and no part of the pattern proceeds without active coordination. The system maintains function, but it does so under full strain, with no remaining buffer capacity to absorb additional variability without altering structure. Patterns persist, but they reflect the maximum extent of compensatory effort, where coordination is sustained only through continuous adjustment.

As saturation is reached, the nature of compensation begins to shift. Rather than stabilizing patterns, it increasingly serves to delay the progression of instability. Threshold crossings become more frequent, as the system operates closer to its structural limits. The magnitude of variability increases, and patterns no longer return to their prior structure even with continued adjustment. The system remains functional, but its behavior reflects a transition from maintaining coherence to managing disruption.

Throughout this progression, baseline drift continues to interact with compensation. Adjustments that sustain function under changing conditions contribute to shifts in the system's reference point, embedding compensation into the baseline itself. The system adapts to this new reference, aligning its responses relative to the current state rather than a prior configuration. Compensation therefore operates not only as a response to variability but as a contributor to the evolving state of the system.

The presence of multiple variability drivers further intensifies this interaction. Input-driven cycles introduce external variability, while stress-driven signals add internally generated variation that may overlap or interrupt existing patterns. Compensation must integrate these sources, aligning responses across signals that may differ in timing and structure. The complexity of this integration increases the demand on coordination, as the system must reconcile competing influences within a single continuous pattern.

Perception reflects the consequences of this increasing demand, though not always directly. Patterns may appear stable, with recognizable transitions and consistent structure, while the effort required to maintain that structure increases internally. As precision declines and adjustments become less aligned, perceptual changes may begin to emerge, reflecting the altered structure of variability. However, the relationship between perception and underlying strain is not always direct, and a disconnect can develop where the system's internal condition is not fully represented in the external signal.

Over longer time horizons, compensation becomes embedded within the system's behavior. Adjusted patterns stabilize around the new conditions created by repeated compensation, forming a baseline that incorporates the history of adjustment. The system adapts to this state, producing patterns that reflect both current variability and accumulated responses. Compensation is no longer a distinct mechanism but part of the continuous structure through which the system operates.

Reintegration of compensation into the broader system reveals its role as a central integrative process. Threshold interactions initiate the need for adjustment, sequencing shapes how variability unfolds, insulin coordination directs response dynamics, baseline drift modifies the reference point, and internal signals introduce additional variation. Compensation connects these mechanisms, ensuring that the system remains coordinated as these factors interact across time. It is not an external addition but an intrinsic feature of continuous system behavior.

This continuity, however, is not without limits. As compensation extends across cycles and accumulates across time, the system approaches conditions where adjustment alone can no longer sustain coherent structure. The progression from precise coordination to layered adjustment to saturation reflects the gradual approach toward these limits. Beyond this point, the system transitions into patterns that reflect structural fatigue, where coordination can no longer fully contain variability, leading into the next stage of system behavior where breakdown and resilience limits become the defining mechanisms.

Amplitude vs Stability Mechanism

Amplitude and stability describe two different dimensions of system behavior that often appear linked but operate independently. Amplitude refers to the scale of movement—the height of rises, the depth of declines, the overall magnitude of variability within a cycle. Stability refers to the consistency with which that movement is organized across time—the repeatability of patterns, the proportional relationships between phases, and the coherence of sequencing from one cycle to the next. The Amplitude vs Stability Mechanism separates these dimensions, establishing that the size of movement does not determine the quality of structure. Large variability can unfold within a stable framework, and minimal variability can occur within patterns that lack coherence. Stability is not defined by how much the system moves, but by whether that movement remains controlled and repeatable across cycles.

The distinction is often obscured by visual interpretation. Smaller movements tend to appear controlled, while larger movements draw attention and are more readily associated with disruption. A narrow range of fluctuation may be perceived as stable simply because it appears contained, while broader oscillations are interpreted as unstable due to their scale. This interpretation reflects a bias toward magnitude as a proxy for structure, where what is seen at a glance is taken as indicative of system behavior. The mechanism challenges this bias by showing that visual compression or expansion of movement does not reveal how that movement is organized across time.

Structure, rather than magnitude, defines stability. A stable pattern maintains consistent relationships between its phases across repeated cycles. The rise follows a similar trajectory relative to baseline, the peak occurs within a predictable range of timing and proportion, the decline aligns with the preceding rise, and the return to baseline completes the sequence in a coherent manner. These relationships persist across cycles, allowing the pattern to be recognized as a repeating structure rather than a series of unrelated events. Stability is therefore expressed through pattern coherence, where variability is organized into a form that can be observed and anticipated through repetition.

High-amplitude patterns can exhibit this coherence even when variability is large. Within a single cycle, the rise may extend significantly from baseline, reaching a pronounced peak before transitioning into a decline that mirrors the scale of the ascent. The magnitude of movement is substantial, but the relationships between phases remain aligned. The timing of transitions follows a consistent sequence, and the proportionality between rise and decline is preserved. The system expresses large variability, but it does so within a structured pattern that maintains internal consistency.

Across multiple cycles, this structure becomes more evident. Each cycle begins from a similar reference point relative to baseline, progresses through a rise that follows a recognizable trajectory, reaches a peak within a comparable range, and transitions into a decline that aligns with the preceding movement. The duration of cycles remains consistent, and the sequencing between them preserves separation and order. Even as amplitude remains high, the pattern repeats with sufficient consistency to be identified as stable. The system maintains alignment across cycles, organizing large variability into a coherent sequence.

This multi-cycle consistency reinforces that stability is not dependent on minimizing movement. The system accommodates significant variation in magnitude while preserving the relationships that define pattern structure. The large amplitude does not disrupt stability because it is integrated into a repeatable framework. Each cycle contributes to the reinforcement of that framework, allowing the pattern to persist across time despite the scale of variability.

In contrast, low-amplitude patterns can lack this coherence even when movement appears minimal. Within a single cycle, the rise may be modest, the peak limited in scale, and the decline relatively shallow. However, the relationships between these phases may not align consistently. The timing of the rise may vary from one cycle to the next, the peak may occur at irregular intervals, and the decline may not proportionally reflect the ascent. The pattern appears contained in magnitude but lacks a consistent structure that can be recognized across cycles.

When observed across multiple cycles, this lack of structure becomes more pronounced. Each cycle may differ in duration, with some extending longer than others or occurring in irregular intervals. The sequencing between cycles may shift, with overlapping or extended gaps that disrupt continuity. The proportional relationships between phases vary, preventing the formation of a repeatable pattern. The system expresses low variability in magnitude but does not organize that variability into a coherent structure. The result is instability, defined not by the size of movement but by the inconsistency of its organization.

This contrast between high-amplitude stability and low-amplitude instability illustrates the independence of these dimensions. One pattern exhibits large variability within a structured framework, while the other exhibits minimal variability within an irregular framework. The difference lies in how the system organizes movement, not in how much movement occurs.

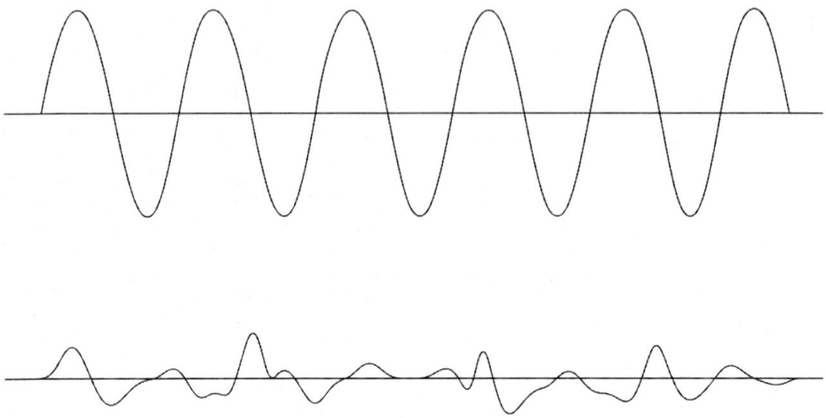

[Figure 16: High amplitude stable vs low amplitude unstable comparison]

The comparison highlights the distinction. In the high-amplitude stable pattern, cycles repeat with consistent shape, timing, and proportional transitions despite large movement. In the low-amplitude unstable pattern, cycles vary in shape and timing, with inconsistent transitions despite small movement. The visual difference in magnitude does not correspond to structural stability, demonstrating that amplitude and stability must be evaluated separately.

The distinction becomes clear only when amplitude and structure are evaluated separately. High-amplitude patterns can maintain precise timing, sequencing, and proportional transitions across cycles, while low-amplitude patterns can lack these relationships entirely. The magnitude of movement does not determine stability. The consistency of structure does.

Understanding stability as controlled variability reframes the role of movement within the system. Stability does not require suppression of variation but organization of that variation into a coherent structure. The system allows movement to occur, but it regulates how that movement unfolds across time. Variability becomes an expression of coordinated behavior rather than a sign of disruption. Stability is achieved not by eliminating change but by maintaining the relationships that define pattern structure.

Instability, in this framework, represents a loss of control over these relationships. It is not characterized by excessive movement alone but by the breakdown of consistency between phases. When amplitude varies unpredictably, timing shifts without pattern, sequencing becomes irregular, and transitions lose coherence, the system exhibits instability regardless of the magnitude of variability. The defining feature is the absence of repeatable structure, where patterns cannot be recognized across cycles due to inconsistency in their organization.

The interaction between compensation and amplitude further illustrates this distinction. Compensation may reduce the visible scale of variability, compressing amplitude without restoring structural alignment, producing patterns that appear stable while remaining irregular beneath the surface. However, this reduction in magnitude does not guarantee stability. If the underlying relationships between phases remain misaligned, the pattern may appear calmer while still lacking structural consistency. Compensation can therefore create the appearance of stability by limiting amplitude, while the organization of variability remains irregular.

Baseline drift introduces another layer to this interaction. As the reference point shifts over time, the relative scale of variability may decrease, reducing visible amplitude. This reduction does not restore structural coherence. The system may move less, but it does not become more stable if timing, sequencing, and transitions remain inconsistent. The position changes, but the organization of variability does not.

Perception often reflects these visual cues, leading to misinterpretation of system behavior. Large movements are more likely to be perceived as disruptive, while smaller movements are perceived as controlled. This perceptual bias aligns with amplitude rather than structure, creating an illusion where high variability appears unstable and low variability appears stable. Perception responds to magnitude, not structure, and therefore often misrepresents stability. However, perception does not capture the relationships between phases across cycles. It responds to the immediate scale of change rather than the consistency of pattern organization.

This perceptual illusion has significant implications for pattern recognition. If recognition is guided by magnitude rather than structure, patterns may be misclassified. A high-amplitude stable pattern may be interpreted as unstable due to its scale, while a low-amplitude unstable pattern may be interpreted as stable due to its containment. Accurate recognition requires prioritizing repeatability and coherence over visual appearance, focusing on how patterns unfold across time rather than how they appear within a single cycle.

Over longer time horizons, the divergence between these pattern types becomes more pronounced. High-amplitude stable patterns reinforce their structure through repetition, preserving consistent relationships between phases across cycles. Low-amplitude unstable patterns do not stabilize over time. They fragment further as inconsistency compounds across cycles, reducing coherence rather than restoring it.

Reintegration of amplitude and stability into the broader system highlights their interaction without conflation. Amplitude defines the scale of variability, while stability defines the organization of that variability. Both dimensions contribute to the overall behavior of the system, but they do so independently. A complete understanding of system behavior requires observing both, recognizing that magnitude alone does not determine structure, and that structure persists or degrades based on the consistency of relationships across time.

As this distinction becomes clear, the focus shifts toward how the system maintains or loses control over pattern structure under varying conditions. Stability, defined as structural coherence, depends on the system's ability to coordinate its mechanisms across cycles. When this coordination is preserved, patterns remain stable regardless of amplitude. When coordination is lost, patterns become unstable regardless of magnitude. This transition from coherence to loss of control introduces the next layer of system behavior, where the durability of patterns under strain and the limits of structural resilience become the defining mechanisms.

Integration Mechanism

Glucose appears as something that can be managed because it is visible, measurable, and responsive to changing conditions. This visibility creates a structural illusion of control, where the signal is mistaken for the mechanism that produces it. The Integration Mechanism separates this illusion from system behavior by defining glucose as an output rather than a controller. The distinction is foundational. Output reflects what the system is doing. Control resides in the interacting processes that generate that output. Glucose does not direct those processes. It expresses their combined behavior.

The signal therefore cannot be treated as an independent variable. It does not initiate movement, coordinate response, resolve variation, or organize cycles. It reflects the outcome of those actions. When glucose rises, it does so because upstream processes have created conditions that produce elevation. When it declines, it reflects coordinated response and clearance. The signal tracks behavior across time, but it does not determine that behavior. The difference between signal and mechanism is not semantic. It defines where explanation must occur.

Multiple subsystems contribute to the formation of this output. External input introduces material for elevation and shapes the initial movement of the signal. Insulin coordination regulates how that material is distributed, taken up, and stored, shaping both rise and decline. Clearance processes determine how efficiently elevation resolves. Internal signaling introduces variability independent of input, adding elevation or disruption from within the system. Sequencing organizes how cycles interact across time, determining whether patterns remain distinct or overlap. Baseline state defines the reference position from which all movement begins. Each of these subsystems contributes to the final pattern, but none acts as a singular controller.

These subsystems must be understood not only as contributors but as interacting components. Their effects do not stack independently in additive fashion. They modify one another. The same input produces different patterns depending on baseline state because elevation begins from a different reference. Internal signaling can alter sequencing by introducing elevation between cycles, changing how subsequent patterns unfold. Compensation modifies how thresholds are approached, redistributing adjustment across timing, magnitude, and duration. Each subsystem operates within the context created by the others, producing a pattern that reflects interaction rather than isolated influence.

This interaction leads to compression at the level of the observable signal. Multiple sources of variability converge into a single curve. The glucose pattern does not display separate traces for input, insulin activity, internal signaling, or baseline positioning. It integrates them into one output. This compression is what makes the signal interpretable and ambiguous at the same time. It provides a unified representation of system behavior while concealing the individual contributions that produced it. The curve shows what happened but not how it happened.

Because the signal is compressed, interpretation becomes structurally difficult. The same visible pattern can arise from different combinations of drivers. A rise may reflect recent input, internal signaling, or a combination of both. A delayed decline may reflect altered clearance, baseline positioning, or compensation effects. The signal does not differentiate these sources. Interpretation therefore depends on recognizing that visible structure reflects interaction rather than origin.

No mechanism operates in isolation within this structure. Input does not act independently of baseline or sequencing. Insulin response does not function independently of amplitude or duration. Internal signaling does not exist separately from compensation or threshold interaction. Each mechanism modifies and is modified by others. The same input can produce different outputs depending on system state. The same glucose level can correspond to different underlying conditions depending on how the pattern was formed. These examples reinforce that output reflects interaction rather than isolated causation.

Pattern formation emerges from this integration. Within a single cycle, the rise, peak, decline, and return are shaped by multiple drivers acting simultaneously. Input may initiate the rise, but its rate and magnitude depend on absorption, baseline position, and hormonal coordination. Decline reflects insulin response, clearance capacity, and sequencing pressure from surrounding cycles. Even within one cycle, no phase is produced by a single mechanism. Each phase is the product of coordinated interaction.

Across multiple cycles, integration becomes more dynamic. The dominance of specific drivers shifts over time. One cycle may be primarily influenced by input structure, while another is shaped more strongly by internal signaling or incomplete clearance. Sequencing determines how these cycles interact, whether they remain separate or overlap. Compensation modifies responses to maintain continuity, while baseline drift shifts the reference point for future cycles. The system does not repeat identical cycles. It produces evolving patterns that reflect the changing interaction of its components.

Reduction attempts to simplify this complexity by attributing patterns to single causes. This approach feels correct because it aligns with the visibility of the signal. A rise is observed, and a cause is assigned. The simplicity of this interpretation creates clarity, but it removes the interactions that produced the pattern. It fails because the signal is not generated by a single driver. The apparent clarity of reduction is therefore an artifact of compression, not a reflection of system behavior. Integrated reality is more complex because it includes all contributing mechanisms simultaneously.

Multi-cycle integration reveals this complexity more clearly. Over time, patterns accumulate the effects of prior cycles. Incomplete clearance carries forward into subsequent rises. Baseline drift alters the starting point of each cycle. Compensation redistributes adjustment across phases, affecting how patterns evolve. Internal signaling may introduce variability independent of input, intersecting with ongoing cycles. Each cycle reflects not only current conditions but the accumulated influence of previous interactions. The system operates as a continuous sequence rather than a series of independent events.

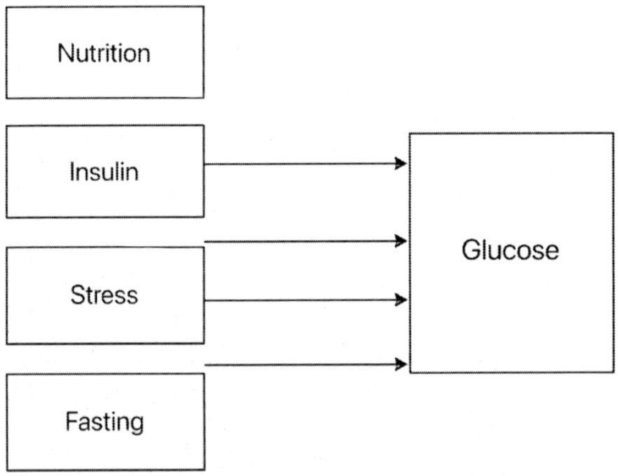

[Figure 17: Multi-system inputs feeding glucose output]

Perception follows this integrated output without revealing its internal structure. Energy, focus, and physical state reflect the visible pattern but do not expose the mechanisms that produced it. The experience corresponds to the signal, not to its sources. This creates a gap between perception and system reality. The same perceptual state may arise from different underlying interactions. A similar rise in perceived energy may result from input-driven elevation in one context and stress-driven elevation in another. The experience is consistent, but the system processes are not.

This gap contributes to interpretive error. Because perception tracks the output, it reinforces the idea that the visible signal represents a singular process. The underlying integration remains hidden, making it difficult to distinguish between different drivers of the same pattern. The system's complexity is compressed into a single experience, further reinforcing the illusion that the signal is self-explanatory.

Stability emerges from integration when subsystems remain aligned across cycles. Alignment does not require uniformity in magnitude but consistency in relationships between phases. When input, response, clearance, sequencing, and baseline state operate in coordination, patterns become repeatable and coherent. Stability is therefore a property of integration rather than a feature of any individual mechanism. It develops gradually as subsystems align across time.

Misalignment also emerges gradually. Partial misalignment may produce patterns that appear coherent but contain subtle inconsistencies in timing, amplitude, or sequencing. These inconsistencies may accumulate across cycles, leading to more visible irregularity. Instability reflects a breakdown in coordinated integration, where relationships between subsystems no longer produce consistent patterns. The transition from stability to instability is therefore a shift in integration quality rather than a change in any single variable.

Compensation operates across this integrated structure, distributing adjustment across multiple subsystems. When variability challenges coordination, compensation modifies timing, magnitude, and sequencing to preserve pattern coherence. These adjustments are not localized. They affect the entire system, altering how subsystems interact. Compensation can maintain visible stability by redistributing strain across mechanisms, masking underlying misalignment. This masking increases coordination demand, as more processes must adjust simultaneously to preserve the pattern.

Thresholds represent boundaries of integrated system behavior rather than limits of a single variable. A threshold is crossed when the combined effect of multiple mechanisms exceeds the system's capacity to maintain coordinated structure. This may involve amplitude, duration, sequencing, baseline drift, or internal signaling, but the resulting shift reflects system-wide change. Thresholds therefore emerge from integration, marking points where coordination can no longer contain variability.

Over longer time horizons, integration accumulates into system state. Repeated interactions between subsystems reshape baseline, alter response dynamics, and influence how patterns evolve. The system retains the effects of prior behavior, forming a continuous record of integrated activity. This accumulation defines how future cycles unfold, as each pattern begins from conditions shaped by previous interactions. Integration is therefore both immediate and historical, operating across cycles and across time.

The full system can be understood as a unified interaction of all mechanisms. Input introduces variability. Amplitude defines scale. Insulin coordinates response. Clearance resolves movement. Sequencing organizes cycles. Baseline drift shifts the reference point. Internal signaling adds variability. Compensation maintains function under strain. Thresholds define limits of coordination. Stability reflects pattern coherence. Each mechanism contributes to the output, and none operates independently.

Glucose is the visible result of this entire structure. It is the compressed expression of system behavior, integrating multiple drivers into a single observable pattern. It cannot be interpreted in isolation because it reflects interaction, not a single process.

When glucose is understood as output, interpretation shifts upstream. The signal remains central because it is observable, but it no longer serves as the sole focus of explanation. Its movement directs attention toward the processes that produced it. Control, in a structural sense, resides within the coordination of subsystems rather than within the signal itself. This shift establishes the basis for examining system-level regulation, where the focus moves from observing output to understanding how coordinated interaction shapes that output across time.

Context Dependency Mechanism

Patterns appear self-explanatory because they are visually coherent. A rise followed by a decline, a repeated sequence of cycles, or a stable-looking curve creates an immediate sense that meaning is embedded in the structure itself. The signal presents as if it contains its own explanation. This creates an illusion of certainty, where recognition of pattern is mistaken for understanding of cause. The Context Dependency Mechanism establishes that this certainty is structural, not analytical. The same observable pattern can emerge from different configurations of the system, and therefore cannot be interpreted without reference to the conditions in which it occurs.

Pattern structure alone is incomplete. It reveals the organization of variability but not the origin of that variability. A curve may show a rapid rise and a gradual decline, or a series of consistent cycles, but these features do not identify which mechanisms produced them. Input-driven elevation, internally generated signaling, compensation adjustments, and baseline positioning can all produce similar visible shapes. The pattern communicates that something has occurred, but it does not specify why it occurred. Interpretation requires moving beyond the structure of the signal into the context that surrounds it.

Multiple sources contribute to pattern formation, and these sources can generate overlapping outputs. Input introduces external variability, shaping rises and initiating cycles. Internal signaling can generate elevation without input, creating patterns that resemble externally driven responses.

Compensation can reshape transitions, producing patterns that appear stable while reflecting underlying adjustment. These drivers do not produce separate signals. They converge into the same observable output, creating patterns that may look similar while arising from different origins. The integration established previously ensures that the signal reflects combined influence rather than isolated mechanisms.

Baseline context modifies how any pattern is expressed and therefore how it must be interpreted. The same rise occurring from a lower baseline represents a different system condition than the same rise occurring from a higher baseline. Amplitude, when viewed without reference to baseline, can appear identical while representing different degrees of deviation from the system's reference state. Baseline determines the relative position of variability, altering how movement relates to system capacity. Without this context, amplitude becomes a misleading measure, as it obscures the relationship between movement and underlying state.

Sequencing context further alters meaning. A single curve viewed in isolation may appear complete, but its significance changes when placed within a sequence of cycles. A rise occurring after full resolution carries different implications than a rise occurring on top of incomplete clearance. Overlapping cycles, compressed spacing, and extended durations all influence how patterns interact across time. The same structural shape may represent a distinct, resolved cycle in one context and a stacked, unresolved sequence in another. Interpretation therefore depends on where a pattern sits within the sequence, not just on the pattern itself.

Temporal context determines whether a pattern represents an isolated event or part of a broader structure. A single cycle cannot define system behavior because variability is continuous and accumulative. Patterns become interpretable only when observed across repeated cycles, where consistency or inconsistency can be established. Short-term observation may capture structure, but it cannot reveal whether that structure persists. Long-horizon observation shows how patterns repeat, shift, or degrade, providing the context necessary to distinguish transient variation from embedded behavior.

Compensation introduces another layer of context that can obscure interpretation. A pattern may appear stable because compensation is actively maintaining its structure. The visible signal can remain coherent while underlying processes adjust continuously to preserve that coherence. In one context, a pattern may reflect natural alignment between subsystems. In another, it may reflect sustained compensatory effort that masks underlying instability. The same visible structure therefore carries different meaning depending on whether it arises from direct coordination or from distributed adjustment.

Threshold context defines whether variability remains within the system's capacity or extends beyond it. The same amplitude may fall within tolerated bounds in one context and exceed them in another. This depends on baseline state, prior cycles, sequencing, and overall system condition. Thresholds are not fixed properties of the signal but emergent properties of system behavior. A pattern that appears similar in magnitude may represent stable variability in one case and threshold crossing in another. Without contextual alignment, the distinction cannot be made.

Perceptual context reinforces the limitations of interpretation based on the signal alone. Experience follows the output, not the mechanisms that produce it. Similar perceptual states can arise from different underlying patterns, and different perceptual states can arise from similar patterns depending on system conditions. The experience of energy, focus, or fatigue reflects the integrated output but does not reveal how that output was generated. This further complicates interpretation, as perception aligns with the visible signal rather than with its contributing processes.

Reduction error emerges when patterns are interpreted without context. The compression of multiple drivers into a single signal encourages the assignment of single causes to observed patterns. A rise may be attributed entirely to input, ignoring internal signaling or baseline conditions. A stable-looking pattern may be assumed to reflect structural stability, ignoring compensation or sequencing effects. Reduction simplifies interpretation but removes the interactions that define system behavior. It creates clarity by discarding context, leading to conclusions that align with the visible signal but not with the processes that produced it.

This misinterpretation is not a minor analytical error. It produces structurally incorrect conclusions about system behavior. When context is removed, the same pattern may be assigned conflicting meanings across different observations, creating inconsistency in interpretation rather than clarity. A rise attributed to input in one instance and to internal signaling in another, without contextual distinction, leads to a fragmented understanding of the system. Over time, this fragmentation prevents accurate recognition of pattern structure, because interpretation becomes detached from the conditions that define it. The system remains coherent, but the interpretation applied to it does not.

The Context Dependency Mechanism defines interpretation as a layered process that requires alignment across multiple contextual dimensions. Baseline context establishes the reference point from which movement is evaluated. Sequencing context determines how patterns interact across cycles. Temporal context defines whether patterns persist or change across time. Driver context identifies the potential sources of variability, whether external or internal. Compensation context reveals whether visible structure reflects direct coordination or active adjustment. Interpretation emerges only when these layers are considered together. Without this alignment, the same pattern can produce multiple conflicting interpretations, none of which reliably reflect system behavior.

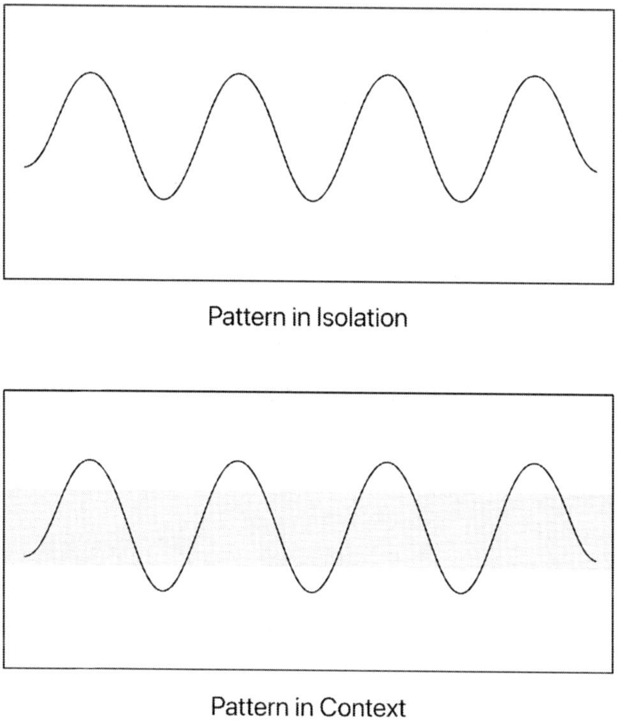

[Figure 18: Same curve, different contextual labels]

The figure illustrates how identical curves can be assigned different meanings depending on context. A single pattern may be labeled as a post-input rise, a stress-driven elevation, a compensated pattern, or a response occurring on an elevated baseline. The structure remains the same, but interpretation diverges based on surrounding conditions. The signal alone does not determine meaning. Context defines it.

Integration with the broader system framework reinforces this dependency. Pattern recognition identifies structure across cycles, distinguishing stable from unstable patterns based on repeatability. Integration establishes that the signal is a product of multiple interacting mechanisms. Thresholds define the limits within which variability can be organized. Compensation maintains visible coherence under changing conditions. Each of these mechanisms contributes to how patterns form and how they must be interpreted. Context links them together, ensuring that interpretation reflects their interaction rather than isolating one dimension.

Over longer horizons, context becomes increasingly important as patterns evolve. Repeated interactions between mechanisms reshape baseline, alter sequencing, and influence compensation demand. The same structural pattern observed at different points in this evolution may represent different system states. Early in a sequence, a pattern may reflect direct coordination. Later, it may reflect accumulated adjustment or drift. Context therefore extends across time, integrating current observation with prior system behavior.

Interpretation, when aligned with context, becomes a system-level observation rather than an event-based conclusion. It recognizes that the signal reflects interaction and that meaning cannot be extracted from structure alone. The visible pattern remains essential because it provides access to system behavior, but it does not provide a complete explanation. It points toward the processes that must be considered to understand that behavior.

The Context Dependency Mechanism establishes a constraint on interpretation. The signal alone is insufficient because it is a compressed representation of system activity. Structure can be recognized, but meaning cannot be assigned without reference to the conditions in which that structure occurs. Context expands the signal, linking visible patterns to the interacting mechanisms and prior states that produced them. Without this expansion, interpretation remains incomplete. This shifts analysis upstream, where system-level coordination must be examined to understand how patterns form and how they should be interpreted across time.

Stability Framing Mechanism

Glucose presents as an accessible point of focus because it is visible, measurable, and responsive to changing conditions. This visibility creates the impression that it is actionable, that movement in the signal can be directly influenced in isolation. The Stability Framing Mechanism begins by dissolving this impression. What appears actionable is not a point of control but a point of observation. The signal is visible because it is the output of the system, not because it is the mechanism that governs the system. The distinction is structural. Signal does not define target, and visibility does not confer control.

A signal reflects system state. It emerges from the interaction of multiple processes operating across time. A target, by contrast, implies that direct manipulation of a single variable can produce a desired outcome. Glucose does not function as a target because it is not a singular variable acting independently. It is the compressed output of integrated subsystems, each contributing to the final pattern. To treat glucose as a target is to treat the output as if it were the source, collapsing the system into a single dimension that cannot account for its own behavior.

This collapse is why control framing fails. Direct control assumes that a change in the signal can be produced by adjusting the signal itself. It assumes that the system responds in a linear and isolated manner, where cause and effect can be traced through a single pathway. The metabolic system does not operate in this way. It is multi-variable and continuously interacting, with each mechanism influencing and being influenced by others. Input, insulin coordination, clearance, sequencing, baseline positioning, internal signaling, compensation, and thresholds all participate in shaping the signal. Attempting to control the output directly disregards these interactions and therefore misaligns with the structure of the system.

Stability does not emerge from suppressing the signal or constraining its movement. It emerges from coordination across mechanisms. When subsystems align in their timing, magnitude, sequencing, and interaction, patterns become coherent and repeatable across cycles. This coherence defines stability. It is not the absence of variability but the organization of variability into a structure that persists over time. Movement remains present, but it is structured rather than erratic. Stability is therefore an emergent property, not a condition imposed on the system.

The distinction between control and stability clarifies the shift in framing required to understand system behavior. Control focuses on immediate changes in the signal, often within a single cycle. It operates at the level of events, responding to visible movement as if each instance were independent. Stability, by contrast, operates across time. It reflects how patterns maintain coherence across repeated cycles, how relationships between phases persist, and how the system organizes variability into a consistent structure. Control is immediate and localized. Stability is cumulative and distributed.

This distinction extends into the difference between reaction and interpretation. Reaction is an event-based response to the signal. It treats each change as a discrete occurrence, prompting a response that is based on the current value or immediate movement. Reaction does not incorporate context, history, or interaction. Interpretation, by contrast, is pattern-based. It requires recognition of structure across cycles, integration of context from multiple mechanisms, and awareness of how the system has evolved over time. Interpretation does not respond to the signal in isolation. It situates the signal within the system that produced it.

Interpretation therefore becomes the primary interface with the system. It connects the observable signal to the underlying mechanisms that generate it. This connection requires alignment with prior chapters. Integration establishes that the signal is a product of multiple interacting subsystems. Context dependency establishes that the same pattern can have different meanings depending on surrounding conditions. Pattern recognition establishes that structure must be observed across cycles to identify stability or instability. Compensation and thresholds define how the system maintains function and where its limits emerge. Interpretation integrates all of these elements, allowing the signal to be understood as part of a larger system rather than as an isolated variable.

The system can be framed as a loop rather than a linear sequence. Subsystems generate variability through interaction. This variability is expressed as the glucose signal. The signal becomes observable and informs interpretation. Interpretation does not act on the signal directly but refines understanding of system behavior. That understanding returns attention to the system itself, where mechanisms continue to interact and produce subsequent output. The loop is continuous, with the signal functioning as an observation point within a larger cycle of system behavior.

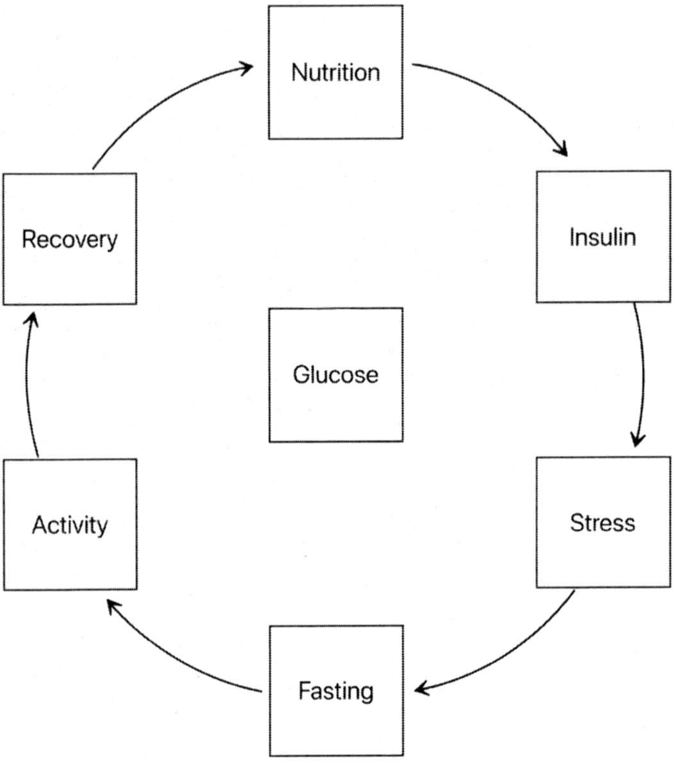

[Figure 19: Glucose as signal within system loop]

Within this loop, glucose occupies the position of output rather than control node. It reflects the state of the system at a given moment, shaped by prior interactions and current conditions. Interpretation sits between the signal and system awareness, translating the visible pattern into an understanding of underlying behavior. The system itself continues to operate through interaction of its mechanisms, independent of the signal's visibility. This framing removes the signal from the role of controller and situates it within the process of observation.

Misinterpretation arises when this loop is collapsed into a direct relationship between signal and action. Treating glucose as a target fragments the system, isolating one variable and attributing to it a level of control it does not possess. This fragmentation leads to misattribution, where patterns are assigned single causes despite being produced by multiple interacting mechanisms. It reinforces reactive thinking, where each change in the signal is treated as an event requiring response rather than as part of a continuous pattern requiring interpretation. The result is a misalignment between how the system operates and how it is understood.

Reintegration of the full system restores this alignment. Input initiates movement but does so within the context of baseline state and sequencing. Insulin coordinates response, shaping how elevation becomes decline and how variability is resolved. Clearance determines how efficiently patterns complete. Sequencing organizes how cycles interact across time. Baseline drift shifts the reference point from which all movement occurs. Internal signaling introduces variability independent of external input. Compensation maintains function under changing conditions, distributing adjustment across mechanisms. Thresholds define the bounds within which coordination can be maintained. Stability reflects the coherence of patterns that emerge from this interaction. Each mechanism contributes to the system's behavior, and the signal reflects their combined influence.

Stability, within this framework, is not defined by flattening the signal or minimizing variability. A flat or minimally variable pattern can still lack coherence if relationships between phases are inconsistent or if patterns do not repeat across cycles. Conversely, patterns with larger amplitude can be stable if they maintain consistent structure, timing, and sequencing. Stability is therefore defined by repeatable structure across time. It reflects how well the system organizes variability, not how much variability is present.

This reframing removes the final interpretive error: treating glucose as something to directly control. The signal cannot be manipulated in isolation because it is not produced in isolation. It is the output of a system that operates through coordinated interaction. Any attempt to control the signal without accounting for this interaction will fail to align with the system's structure. Understanding must therefore precede any notion of influence, and that understanding must remain at the level of the system rather than the signal.

The Stability Framing Mechanism establishes a constraint rather than a method. It defines how the system must be viewed, not how it should be acted upon. Glucose is positioned as a signal within a continuous loop of system behavior. Stability is positioned as an emergent outcome of coordinated interaction across mechanisms. Interpretation is positioned as the primary interface through which the system can be understood. These positions redefine the relationship between observer and system, replacing direct control with structured observation.

As this framing stabilizes, the signal remains central but no longer serves as a target. It functions as a reference point within a broader system of interaction, directing attention toward the processes that generate it. Stability is not imposed on the signal but emerges when system mechanisms align across time. Interpretation provides the only access to that alignment, linking observable output to the structure that produces it. Without this framing, the signal is misread. With it, the system becomes visible.

About the Author

Nathan Reed writes about metabolic stability in systems exposed to load, variability, and time.

His work examines how biological capacity is shaped by energy availability, recovery architecture, sleep amplitude, muscle mass, environmental demand and workload. Rather than treating nutrition, fasting, sleep, and training as isolated tactics, he approaches them as interacting regulatory systems governed by margin, buffering, oscillatory coherence, and constraint dominance.

The Stability framework explores how physiological compression accumulates gradually and how proportional calibration preserves long-horizon resilience across decades. The objective is not peak optimization, but preservation of functional continuity under load.

Glucose Stability is part of a broader metabolic stability ecosystem developed under Perennial Publishing, each volume addressing a distinct vector of biological durability.

The work is designed for readers who prefer structural clarity over motivational rhetoric and long-horizon coherence over short-term intensity.

The Metabolic Stability Series

Metabolic Stability: *A Systems Guide to Blood Sugar, Fasting, and Long-Term Metabolic Health*

(Cornerstone Volume)

Fasting Fundamentals: *A Systems Guide to Metabolic Flexibility, Fat Loss, and Long-Term Health*

(Foundational Volume)

Intermittent Fasting: *Metabolic Timing, Body Composition, and Sustainable Health*

Prolonged Fasting: *A Systems Guide to Duration, Recovery, and Metabolic Limits*

Electrolytes: *A Systems Guide to Hydration, Electrolyte Balance, and Physiological Stability*

Insulin Stability: *Blood Sugar, Insulin Resistance, and Metabolic Health*

Glucose Stability: *How Blood Sugar Fluctuations Affect Energy, Insulin, and Metabolic Health*

Future Titles Include:

Energy Stability

Adaptation and Drift

The Vitality Series:

Sleep and Recovery

Strength and Aging

Hormones & Sexual Health

Part of a Broader Stability Framework

The Metabolic Stability Series is part of a larger stability ecosystem published by Perennial Publishing.

Related domains include:

Money Stability — structural durability in financial systems

Career Stability — positioning, load tolerance, and transition resilience

Mental Stability — cognitive steadiness under pressure and uncertainty

Emotional Stability — regulatory steadiness in environments that do not slow down

Foundations — low-fragility life architecture

Vitality — energy, strength, and biological function across time

Each domain approaches stability as a structural problem: capacity versus load, buffering versus shock, governance versus drift. Together, they form a unified framework for long-horizon durability across the major systems that shape human trajectory.

About Perennial Publishing

Perennial Publishing produces practical systems guides for building stability in an unstable world.

Our books explore the foundations of long-term resilience across health, finance, career, and everyday decision-making. Each title focuses on clear frameworks, durable principles, and practical strategies that help readers build lasting stability rather than short-term fixes.

The Perennial catalog includes series on metabolic health, financial stability, career systems, and personal consistency. While each book stands on its own, they share a common goal: helping individuals create durable systems that support a stable and sustainable life.